Revision Guide

T0076393

Cambridge

International AS and A Level

Biology

Second Edition

Mary Jones

HODDER
EDUCATION
LEARN MORE

Hodder Education, an Hachette UK company, Carmelite House, 50 Victoria Embankment, London EC4Y 0DZ

Orders

Bookpoint Ltd, 130 Park Drive, Milton Park, Abingdon, Oxfordshire OX14 4SE
tel: 01235 827827
fax: 01235 400401
e-mail: education@bookpoint.co.uk
Lines are open 9.00 a.m.–5.00 p.m., Monday to Saturday, with a 24-hour message answering service. You can also order through the Hodder Education website: www.hoddereducation.co.uk

© Mary Jones 2015
ISBN 978-1-4718-2887-4

First printed 2015
Impression number 5 4 3 2 1
Year 2019 2018 2017 2016 2015

All rights reserved; no part of this publication may be reproduced, stored in a retrieval system, or transmitted, in any form or by any means, electronic, mechanical, photocopying, recording or otherwise without either the prior written permission of Hodder Education or a licence permitting restricted copying in the United Kingdom issued by the Copyright Licensing Agency Ltd, Saffron House, 6–10 Kirby Street, London EC1N 8TS.

Cover photo reproduced by permission of Fotolia

Typeset by Greenhill Wood Studios, UK
Printed in Spain

This text has not been through the Cambridge endorsement process.

Hachette UK's policy is to use papers that are natural, renewable and recyclable products and made from wood grown in sustainable forests. The logging and manufacturing processes are expected to conform to the environmental regulations of the country of origin.

Get the most from this book

Everyone has to decide his or her own revision strategy, but it is essential to review your work, learn it and test your understanding. This Revision Guide will help you to do that in a planned way, topic by topic. Use the book as the cornerstone of your revision and don't hesitate to write in it — personalise your notes and check your progress by ticking off each section as you revise.

☑ Tick to track your progress

Use the revision planner on pages 4 and 5 to plan your revision, topic by topic. Tick each box when you have:

- revised and understood a topic
- tested yourself
- practised the exam-style questions

You can also keep track of your revision by ticking off each topic heading in the book. You may find it helpful to add your own notes as you work through each topic.

My revision planner

AS topics

		Revised	Tested	Exam ready
1 Cell structure				
7	Microscopy	☐	☐	☐
10	Cell structure and function	☐	☐	☐
2 Biological molecules				
14	Carbohydrates	☐	☐	☐

Microscopy

Light microscopes and electron microscopes `Revised ☐`

Cells are the basic units from which living organisms are made. Most cells are very small, and their structures can only be seen by using a microscope.

You will use a light microscope during your AS course. Light rays pass through the specimen on a slide and are focused by an objective lens and an eyepiece lens. This produces a magnified image of the specimen on the retina of your eye. Alternatively, the image can be projected onto a screen, or recorded by a

Features to help you succeed

Expert tips

Throughout the book there are tips from the experts on how to maximise your chances.

Typical mistakes

Advice is given on how to avoid the typical mistakes students often make.

Definitions and key words

Clear, concise definitions of essential key terms are provided on the page where they first appear.

Key words from the syllabus are highlighted in bold for you throughout the book.

Exam-style questions

Exam-style questions are provided for AS and A level. Use them to consolidate your revision and practise your exam skills.

Now test yourself

These short, knowledge-based questions provide the first step in testing your learning. Answers are at the back of the book.

Revision activities

These activities will help you to understand each topic in an interactive way.

My revision planner

Countdown to my exams

6–8 weeks to go

- Start by looking at the syllabus — make sure you know exactly what material you need to revise and the style of the examination. Use the revision planner on pages 4 and 5 to familiarise yourself with the topics.
- Organise your notes, making sure you have covered everything on the syllabus. The revision planner will help you to group your notes into topics.
- Work out a realistic revision plan that will allow you time for relaxation. Set aside days and times for all the subjects that you need to study, and stick to your timetable.
- Set yourself sensible targets. Break your revision down into focused sessions of around 40 minutes, divided by breaks. This Revision Guide organises the basic facts into short, memorable sections to make revising easier.

Revised ☐

2–5 weeks to go

- Read through the relevant sections of this book and refer to the expert tips, typical mistakes and key terms. Tick off the topics as you feel confident about them. Highlight those topics you find difficult and look at them again in detail.
- Test your understanding of each topic by working through the 'Now test yourself' questions in the book. Look up the answers at the back of the book.
- Make a note of any problem areas as you revise, and ask your teacher to go over these in class.
- Look at past papers. They are one of the best ways to revise and practise your exam skills. Write or prepare planned answers to the exam-style questions provided in this book. Check your answers with your teacher.
- Try different revision methods. For example, you can make notes using mind maps, spider diagrams or flash cards.
- Track your progress using the revision planner and give yourself a reward when you have achieved your target.

Revised ☐

1 week to go

- Try to fit in at least one more timed practice of an entire past paper and seek feedback from your teacher, comparing your work closely with the mark scheme.
- Check the revision planner to make sure you haven't missed out any topics. Brush up on any areas of difficulty by talking them over with a friend or getting help from your teacher.
- Attend any revision classes put on by your teacher. Remember, he or she is an expert at preparing people for examinations.

Revised ☐

The day before the examination

- Flick through this Revision Guide for useful reminders, for example the expert tips, typical mistakes and key terms.
- Check the time and place of your examination.
- Make sure you have everything you need — extra pens and pencils, tissues, a watch, bottled water, sweets.
- Allow some time to relax and have an early night to ensure you are fresh and alert for the examinations.

Revised ☐

My exams

Paper 1

Date: .... Time: ....

Location:...

Paper 2

Date: .... Time: ....

Location:...

Paper 3

Date: .... Time: ....

Location:...

Paper 4

Date: .... Time: ....

Location:...

Paper 5

Date: .... Time: ....

Location:...

1 Cell structure

Microscopy

Light microscopes and electron microscopes

Revised ☐

Cells are the basic units from which living organisms are made. Most cells are very small, and their structures can only be seen by using a microscope.

You will use a light microscope during your AS course. Light rays pass through the specimen on a slide and are focused by an objective lens and an eyepiece lens. This produces a magnified image of the specimen on the retina of your eye. Alternatively, the image can be projected onto a screen, or recorded by a camera.

An electron microscope uses beams of electrons rather than light rays. The specimen has to be very thin and must be placed in a vacuum, to allow electrons to pass through it. The electrons are focused onto a screen, or onto photographic film, where they form a magnified image of the specimen.

Magnification and resolution

Magnification can be defined as:

$$\text{magnification} = \frac{\text{size of image}}{\text{actual size of object}}$$

This can be rearranged to:

$$\text{size of object} = \frac{\text{size of image}}{\text{magnification}}$$

There is no limit to the amount you can magnify an image. However, the amount of useful magnification depends on the **resolution** of the microscope. This is the ability of the microscope to distinguish two objects as separate from one another. The smaller the objects that can be distinguished, the higher the resolution. Resolution is determined by the wavelength of the rays that are being used to view the specimen. The wavelength of a beam of electrons is much smaller than the wavelength of light. An electron microscope can therefore distinguish between much smaller objects than a light microscope — in other words, an electron microscope has a much higher resolution than a light microscope. We can therefore see much more fine detail of a cell using an electron microscope than using a light microscope.

As cells are very small, we have to use units much smaller than millimetres to measure them. These units are micrometres, **µm**, and nanometres, **nm**.

$1\,\text{mm} = 1 \times 10^{-3}\,\text{m}$

$1\,\mu\text{m} = 1 \times 10^{-6}\,\text{m}$

$1\,\text{nm} = 1 \times 10^{-9}\,\text{m}$

To change mm into µm, multiply by 1000.

> **Magnification** is the size of an image divided by the size of the actual object.
>
> **Resolution** is the size of the smallest objects that can be distinguished.

> **Expert tip**
>
> It is almost always a good idea to convert every measurement to µm when doing magnification calculations.

How to... Calculate magnification

You should be able to work out the real size of an object if you are told how much it has been magnified.

For example, the drawing of a mitochondrion in Figure 1.1 has been magnified 100 000 times.

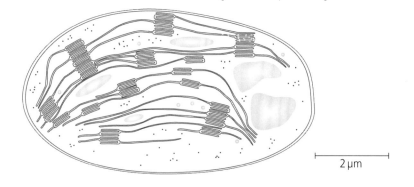

Figure 1.1

*   Use your ruler to measure its length in mm. It is 50 mm long.
*   As it is a very small object, convert this measurement to µm by multiplying by 1000:

    $50 \times 1000 = 50\,000\,\mu m$
*   Substitute into the equation:

    $$\text{actual size of object} = \frac{\text{size of image}}{\text{magnification}}$$

    $$= \frac{50\,000}{100\,000}$$

    $$= 0.5\,\mu m$$

You can also use a scale bar to do a similar calculation for the drawing of a chloroplast in Figure 1.2.

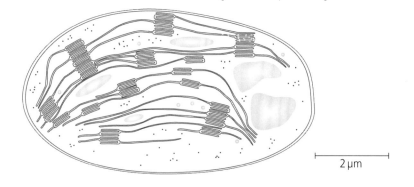

2 µm

Figure 1.2

*   Measure the length of the scale bar.
*   Calculate its magnification using the formula

    $$\text{magnification} = \frac{\text{size of image}}{\text{actual size of object}}$$

    $$= \frac{\text{length of scale bar}}{\text{length the scale bar represents}}$$

    $$= \frac{20\,000}{2}$$

    $$= \times 10\,000$$

*   Measure the length of the image of the chloroplast in mm, and convert to µm. You should find that it is 80 000 µm long.
*   Calculate its real length using the formula

    $$\text{actual size of object} = \frac{\text{size of image}}{\text{magnification}}$$

    $$= \frac{80\,000}{10\,000}$$

    $$= 8\,\mu m$$

Expert tip

Always show every small step in your working when you do a calculation on an exam paper, as there may be marks for this.

Now test yourself

Tested

1   A micrograph shows a chloroplast that measures 79 mm long. The magnification of the micrograph is ×16 000. Calculate the length of the chloroplast in µm.

Answer on p.202

How to... **Measure cells using a graticule**

An eyepiece graticule is a little scale bar that you can place in the eyepiece of your light microscope. When you look down the microscope, you can see the graticule as well as the specimen.

The graticule is marked off in 'graticule units', so you can use the graticule to measure the specimen you are viewing in these graticule units. Just turn the eyepiece so that the graticule scale lies over the object you want to measure. It will look like Figure 1.3.

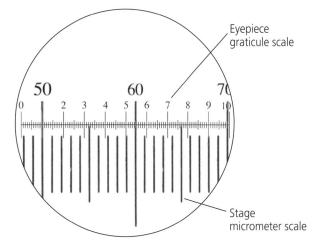

Figure 1.4

- You can see that the 50 mark on the stage micrometer is lined up with the 1.0 mark on the eyepiece graticule. Work along towards the right until you see another two lines that are exactly lined up. There is a good alignment of 68 on the stage micrometer and 9.0 on the eyepiece graticule. So you can say that:

  80 small eyepiece graticule markings = 18 stage micrometer markings

  $= 18 \times 0.01\,mm = 0.18\,mm$

  $= 180\,\mu m$

  so 1 small eyepiece graticule marking $= \dfrac{180}{80} = 2.25\,\mu m$.

Now we can calculate the real width of the plant cell we measured. It was 23 eyepiece graticule units long. So its real width is:

$23 \times 2.25 = 51.75\,\mu m$

If you want to look at something using a different objective lens, you will have to do the calibration of eyepiece graticule units all over again using this lens. Once you have done it, you can save your calibrations for the next time you use the same microscope with the same eyepiece graticule and the same objective lens.

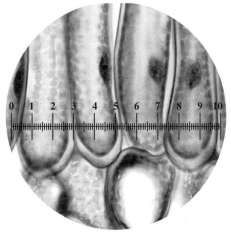

Figure 1.3

We can say that the width of one cell is 23 graticule units.

The graticule units have to be converted to real units, such as mm or μm. This is called **calibration**. To do this, you use a special slide called a **stage micrometer** that is marked off in a tiny scale. There should be information on the slide that tells you the units in which it has been marked. The smallest markings are often 0.01 mm apart — that is, 10 μm apart.

Take the specimen off the stage or the microscope and replace it with the stage micrometer. Focus on it using the same objective lens as you used for viewing the specimen.

Line up the micrometer scale and the eyepiece graticule scale. You can do this by turning the eyepiece, and by moving the micrometer on the stage. Make sure that two large markings on each scale are exactly lined up with each other. You should be able to see something like Figure 1.4.

Now test yourself

Tested

2 A cell measures 84 small eyepiece graticule units.

When a stage micrometer is placed on the stage using the same objective lens, it can be seen that 100 eyepiece graticule units exactly line up with 8 stage micrometer markings. The stage micrometer is marked off in 0.01 mm divisions.

Calculate the size of the cell in μm.

Answer on p.202

Cell structure and function

Figures 1.5 and 1.6 show a typical animal cell and a typical plant cell as seen using an electron microscope.

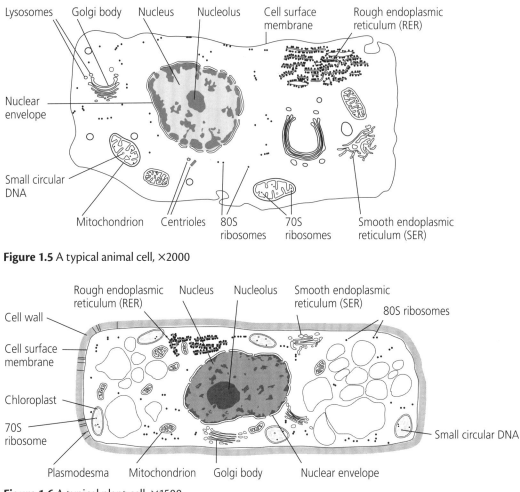

Figure 1.5 A typical animal cell, ×2000

Figure 1.6 A typical plant cell, ×1500

Functions of membrane systems and organelles

Revised

The **cell surface membrane** controls what enters and leaves the cell. Its structure and functions are described in detail on pp. 32–33. There are also many membranes within the cell that help to make different compartments in which different chemical reactions can take place without interfering with one another.

The **nucleus** is surrounded by a pair of membranes, which make up the **nuclear envelope**. The nucleus contains **chromosomes**, each of which contains a very long molecule of DNA. The DNA determines the sequences in which amino acids are linked together in the cytoplasm to form protein molecules. This is described on pp. 44–45.

Within the nucleus there is a darker area (not surrounded by its own membrane) called the **nucleolus**. This is where new ribosomes are made, following a code on part of the DNA.

Ribosomes are small structures made of RNA and protein. They are found free in the cytoplasm, and also attached to **rough endoplasmic reticulum (RER)**. These ribosomes are 80S (that is, larger than those in the mitochondria and chloroplasts). The RER is an extensive network of membranes in the cytoplasm. The membranes enclose small spaces called **cisternae**. Proteins are made on the ribosomes, by linking together amino acids.

If the proteins are to be processed or exported from the cell, the growing chains of amino acids move into the cisternae of the RER as they are made. The cisternae then break off to form little **vesicles** that travel to the **Golgi body**. Here they may be modified, for example by adding carbohydrate groups to them. Vesicles containing the modified proteins break away from the Golgi body and are transported to the cell surface membrane (Figure 1.7), where they are secreted from the cell by exocytosis (p. 38).

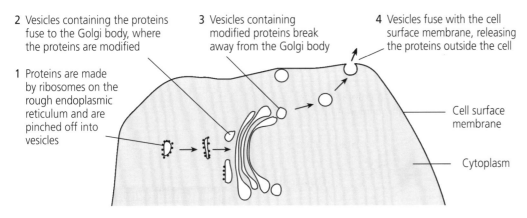

2 Vesicles containing the proteins fuse to the Golgi body, where the proteins are modified

3 Vesicles containing modified proteins break away from the Golgi body

4 Vesicles fuse with the cell surface membrane, releasing the proteins outside the cell

1 Proteins are made by ribosomes on the rough endoplasmic reticulum and are pinched off into vesicles

Cell surface membrane

Cytoplasm

Figure 1.7 The interrelationship between RER and the Golgi body

Smooth endoplasmic reticulum (**SER**) is usually less extensive than RER. It does not have ribosomes attached to it, and the cisternae are usually more flattened than those of the RER. It is involved in the synthesis of steroid hormones and the breakdown of toxins.

Mitochondria have an envelope (two membranes) surrounding them (Figure 1.8). The inner one is folded to form **cristae**. This is where aerobic respiration takes place, producing ATP (see p. 12). The first stage of this process, called the Krebs cycle, takes place in the **matrix**. The final stage, oxidative phosphorylation, takes place on the membranes of the cristae. Mitochondria contain ribosomes that are smaller than those in the cytoplasm (70S) and a small circular molecule of DNA.

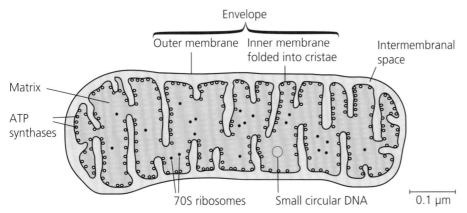

Envelope

Outer membrane   Inner membrane folded into cristae

Intermembranal space

Matrix

ATP synthases

70S ribosomes   Small circular DNA

0.1 µm

Figure 1.8 Longitudinal section through a mitochondrion

Lysosomes are little membrane-bound packages of hydrolytic (digestive) enzymes. They form by breaking off from the Golgi body. They are used to digest bacteria or other cells taken into the cell by phagocytosis, or to break down unwanted or damaged organelles within the cell.

Centrioles are found only in animal cells, not plant cells. They are made of tiny structures called **microtubules**, arranged in a circular pattern. Microtubules are made of a protein called actin. The two centrioles lie at right angles to one another. It is from here that the microtubules are made that form the spindle during cell division in animal cells. Microtubules are also found throughout the cell even when it is not dividing, where they help to form the cytoskeleton, which keeps the cell in shape.

Chloroplasts are found only in some plant cells. Like mitochondria, chloroplasts are surrounded by an envelope made up of two membranes (Figure 1.9). Their background material is called the **stroma**, which contains many paired membranes called **thylakoids**. In places, these form stacks called **grana**. The grana contain chlorophyll, which absorbs energy from sunlight. The first reactions in photosynthesis, called the light dependent reactions and photophosphorylation, take place on the membranes. The final stages, called the Calvin cycle, take place in the stroma. Chloroplasts often contain starch grains, which are storage materials formed from the sugars that are produced in photosynthesis. Like mitochondria, they contain 70S ribosomes and a circular DNA molecule.

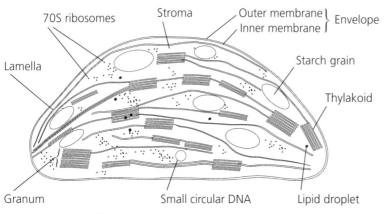

Figure 1.9 Longitudinal section through a chloroplast

Plant cells are always surrounded by a **cell wall** made of cellulose, which is never found around animal cells. The structure of cellulose is described on p. 16–17.

ATP

Cells constantly use energy to maintain the processes that keep them alive. These processes include active transport of substances across membranes (p. 37), movement and the building up of large molecules (such as the synthesis of proteins, p. 44). The immediate source of energy for a cell is adenosine triphosphate, or ATP. ATP can be broken down to adenosine diphosphate (ADP) and a phosphate ion, releasing a small packet of useable energy. Mitochondria make ATP by aerobic respiration (p. 101), while chloroplasts make ATP in the light dependent stage of photosynthesis (p. 109).

Revision activity

Cut out 21 pieces of card, all the same size. On one side of each card write one of the words shown in **bold** on pp. 10–12. On the other side, write no more than six words that describe the named structure. Arrange all the cards with either the name or the description visible, and practise identifying what is on the non-visible side.

Prokaryotic cells

Revised

Prokaryotic cells are found in bacteria and archaea, whereas eukaryotic cells are found in animals, plants, protoctista and fungi. Prokaryotic cells are generally much smaller than eukaryotic cells. The fundamental difference between prokaryotic and eukaryotic cells is that they do not have a nucleus or any other organelles bound by a double membrane (Figure 1.10). Most prokaryotic organisms are unicellular. Eukaryotic organisms (including plants, animals and fungi) are usually multicellular. Prokaryotic and eukaryotic cells are compared in Table 1.1.

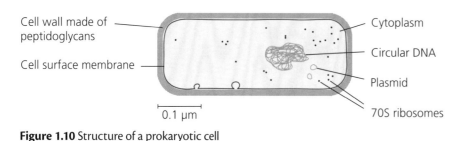

Figure 1.10 Structure of a prokaryotic cell

Table 1.1 Comparison of prokaryotic, animal and plant cells

| Feature | Prokaryotic cells | Eukaryotic cells | |
		Animal cells	Plant cells
Cell surface membrane	Always present	Always present	Always present
Cell wall	Always present; made up of peptidoglycans	Never present	Always present; made up of cellulose
Nucleus and nuclear envelope	Never present	Always present	Always present
Chromosomes	Contain so-called 'bacterial chromosomes' — a circular molecule of DNA not associated with histones (sometimes said to be 'naked' DNA); bacteria may also contain smaller circles of DNA called plasmids	Contain several chromosomes, each made up of a linear DNA molecule associated with histones	Contain several chromosomes, each made up of a linear DNA molecule associated with histones
Mitochondria	Never present	Usually present	Usually present
Chloroplasts	Never present, though some do contain chlorophyll or other photosynthetic pigments	Never present	Sometimes present
Rough and smooth endoplasmic reticulum and Golgi bodies	Never present	Usually present	Usually present
Ribosomes	70S ribosomes present	80S ribosomes present in cytoplasm; 70S ribosomes in mitochondria	80S ribosomes present in cytoplasm; 70S ribosomes in mitochondria and chloroplasts
Centrioles	Never present	Usually present	Never present

Revision activity

● Using a good quality HB pencil, make a large diagram of a prokaryotic cell. Check your diagram against Figure 1.10.

● Now convert the prokaryotic cell to an animal cell, by erasing some structures and adding others. Check your diagram against Figure 1.5.

● Now convert the animal cell to a plant cell, by erasing one structure and adding others. Check your diagram against Figure 1.6.

Viruses

Revised

Viruses are extremely small particles of RNA or DNA surrounded by a protein coat (Figure 1.11). They are non-cellular. Viruses are not able to carry out any of the processes characteristic of living things unless they are inside a living cell, where they 'hijack' the cell's machinery to make copies of themselves. All viruses are, therefore, parasitic. Some viruses cause disease in humans.

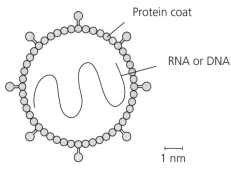

Protein coat

RNA or DNA

1 nm

Figure 1.11 Structure of a virus

Now test yourself

3 Use the scale bar to calculate the diameter of the virus in Figure 1.11.

Then work out how many viruses could line up inside the bacterium shown in Figure 1.10.

Answer on p.202

Tested

2 Biological molecules

Carbohydrates

Carbohydrates are substances whose molecules contain carbon, hydrogen and oxygen atoms, and in which there are approximately twice as many hydrogen atoms as carbon or oxygen atoms.

Monosaccharides and disaccharides

Revised

The simplest carbohydrates are **monosaccharides**. These are sugars. They include glucose, fructose and galactose. These three monosaccharides each have six carbon atoms, so they are also known as hexose sugars. Their molecular formula is $C_6H_{12}O_6$.

Monosaccharide molecules are often in the form of a ring made up of carbon atoms and one oxygen atom. Glucose molecules can take up two different forms, called α-glucose and β-glucose (Figure 2.1).

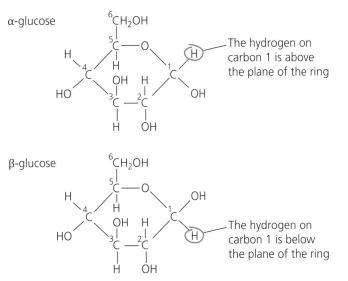

Figure 2.1 Structural formulae of α-glucose and β-glucose molecules

> ### Expert tip
> When you are drawing biological molecules, count the bonds you have drawn on each atom. Each C should have four bonds, each N should have 3, each O should have 2 and each H should have 1.

Two monosaccharides can link together to form a **disaccharide**. For example, two glucose molecules can link to produce **maltose**. The bond that joins them together is called a **glycosidic bond**. As the two monosaccharides react and the glycosidic bond forms, a molecule of water is released. This type of reaction is known as a **condensation reaction**. Different disaccharides can be formed by linking different monosaccharides (Table 2.1 and Figures 2.2 and 2.3).

Table 2.1

Disaccharide	Monosaccharides
Maltose	Glucose + Glucose
Lactose	Glucose + Galactose
Sucrose	Glucose + Fructose

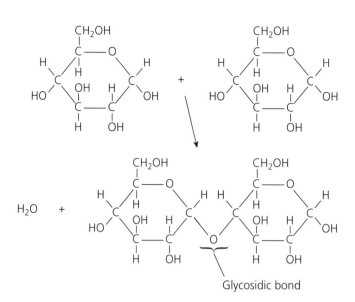

Figure 2.2 Formation of maltose by a condensation reaction

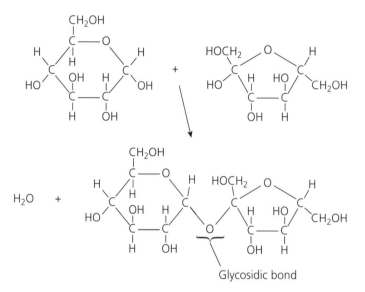

Figure 2.3 Formation of sucrose by a condensation reaction

Disaccharides can be split apart into two monosaccharides by breaking the glycosidic bond. To do this, a molecule of water is added. This is called a **hydrolysis reaction** (Figure 2.4).

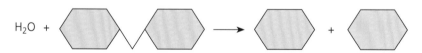

Figure 2.4 Breakdown of a disaccharide by a hydrolysis reaction

Functions of monosaccharides and disaccharides

- Monosaccharides and disaccharides are good sources of energy in living organisms. They can be used in respiration, in which the energy they contain is used to make ATP.
- Because they are soluble, they are the form in which carbohydrates are transported through an organism's body. In animals, glucose is transported dissolved in blood plasma. In plants, sucrose is transported in phloem sap.

All monosaccharides and some disaccharides act as reducing agents, and will reduce blue Benedict's solution to produce an orange-red precipitate. They are called **reducing sugars**. Sucrose is a **non-reducing sugar**.

Expert tip

Remember: carbohydrates are transported as glucose in animals, but as sucrose in plants.

Now test yourself

1 For each of these sugars, state:
   a whether it is a monosaccharide or a disaccharide
   b if a disaccharide, the monosaccharides from which it is formed
   c whether it will give a positive result with the Benedict's test

   fructose   glucose   lactose
   maltose   sucrose

Answer on p.202

Tested

Polysaccharides

These are substances whose molecules contain hundreds or thousands of monosaccharides linked together into long chains. The monosaccharides are **monomers**, and the polysaccharide is a **polymer**. Polysaccharides are **macromolecules**. Because their molecules are so enormous, the majority do not dissolve in water. This makes them good for storing energy (starch and glycogen) or for forming strong structures (cellulose).

Storage polysaccharides

In animals and fungi, the storage polysaccharide is **glycogen**. It is made of α-glucose molecules linked together by glycosidic bonds. Most of the glycosidic bonds are between carbon 1 on one glucose and carbon 4 on the next, so they are called 1–4 links. There are also some 1–6 links, which form branches in the chain (Figure 2.5). When needed, the glycosidic bonds can be hydrolysed by carbohydrase enzymes to form monosaccharides, which can be used in respiration. The branches mean there are many 'ends', which increases the rate at which carbohydrases can hydrolyse the molecules.

> A **monomer** is a relatively small molecule that can be linked to others like it, to form a much longer molecule.
>
> A **polymer** is a molecule made up of many smaller molecules (monomers) linked together in a long line.
>
> A **macromolecule** is a large biological molecule, such as a polysaccharide or protein.

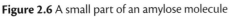

Figure 2.5 A small part of a glycogen molecule

In plants, the storage polysaccharide is **starch**. Starch is a mixture of two substances, **amylose** and **amylopectin**. An amylose molecule is a long chain of α-glucose molecules with 1–4 links. It coils up into a spiral, making it very compact. The spiral is held in shape by hydrogen bonds (see p. 23) between small charges on some of the hydrogen and oxygen atoms in the glucose units (Figure 2.6). An amylopectin molecule is similar to glycogen.

(see p. 23)

> **Typical mistake**
>
> It's important to remember that you do not find starch in animal cells, nor glycogen in plant cells.

Amylose contains chains of α 1–4 linked glucose

The glucose molecules in a chain all have the same orientation

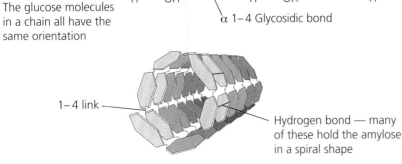

α 1–4 Glycosidic bond

1–4 link

Hydrogen bond — many of these hold the amylose in a spiral shape

Figure 2.6 A small part of an amylose molecule

Structural polysaccharides

Plant cell walls contain the polysaccharide **cellulose**. Like amylose, this is made of many glucose molecules linked by glycosidic bonds between carbon 1 and carbon 4 (Figure 2.7). However, in cellulose the glucose molecules are in the β form. This means that adjacent glucose molecules in the chain are upside-down in relation to one another. The chain stays straight, rather than spiralling.

> A **monosaccharide** is a single sugar unit, with the formula $C_nH_{2n}O_n$, e.g. glucose.
>
> A **disaccharide** is a substance made of two monosaccharide molecules linked by a glycosidic bond, e.g. maltose.
>
> A **polysaccharide** is a substance made of many monosaccharide molecules linked in a long chain, e.g. glycogen.

Hydrogen bonds form between different chains (Figure 2.8). This causes the chains to associate into bundles called **microfibrils**.

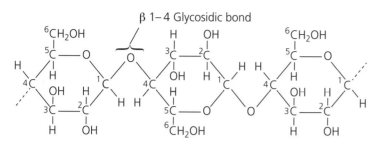

The glucose molecules in a chain alternate in their orientation (look at the numbered carbon atoms)

Figure 2.7 Cellulose contains chains of β 1–4 linked glucose

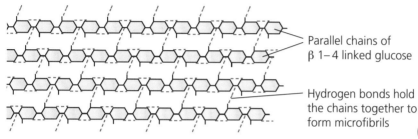

Parallel chains of β 1–4 linked glucose

Hydrogen bonds hold the chains together to form microfibrils

Figure 2.8 Cellulose molecules

The resulting microfibrils are very strong. This makes cellulose an excellent material for plant cell walls, because it will not break easily if the plant cell swells as it absorbs water. The microfibrils are also very difficult to digest, because few organisms have an enzyme that can break the β 1–4 glycosidic bonds.

> **Now test yourself**
>
> 2 Explain why cellulose works well as a structural polysaccharide, whereas starch does not.
>
> **Answer on p.202**
>
> Tested

Tests for carbohydrates

Revised

Reducing sugar

Add Benedict's reagent and heat. An orange-red precipitate indicates the presence of reducing sugar. If standard volumes of the unknown solutions and excess Benedict's reagent are used, the mass of precipitate or intensity of the orange-red colour indicates the concentration of the solution. This can be matched against colour standards, prepared using reducing sugar solutions of known concentration. (See p. 78 for an explanation of how to make solutions of known concentration.)

Non-reducing sugar

This test should only be done on solutions known not to contain reducing sugars. Some disaccharides, for example sucrose, are non-reducing sugars. Hydrolyse (break the glycosidic bond) by heating with dilute HCl, then neutralise with sodium hydrogencarbonate. Then carry out the test for reducing sugar.

Starch

Add iodine in potassium iodide solution. A blue-black colour indicates the presence of starch.

> **Now test yourself**
>
> 3 Imagine you have been given a sugar solution. You know that it contains reducing sugar, but you want to know if it also contains non-reducing sugar. How could you find out?
>
> **Answer on p.202**
>
> Tested

Lipids

Lipids, like carbohydrates, also contain carbon, hydrogen and oxygen, but there is a much smaller proportion of oxygen. Lipids include triglycerides and phospholipids. All lipids are insoluble in water.

Triglycerides

A triglyceride molecule is made of a 'backbone' of glycerol, to which three fatty acids are attached by **ester bonds** (Figure 2.9).

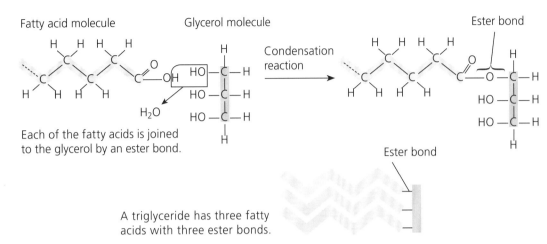

Each of the fatty acids is joined to the glycerol by an ester bond.

A triglyceride has three fatty acids with three ester bonds.

Figure 2.9 The formation of a triglyceride molecule

Fatty acids have long chains made of carbon and hydrogen atoms. Each carbon atom has four bonds. Usually, two of these bonds are attached to other carbon atoms, and the other two to hydrogen atoms. In some cases, however, there may be only one hydrogen atom attached. This leaves the carbon atom with a 'spare' bond, which attaches to the next-door carbon atom (which also has one less hydrogen bonded to it), forming a double bond. Fatty acids with one or more carbon–carbon double bonds are called unsaturated fatty acids, because they do not contain quite as much hydrogen as they could. Fatty acids with no double bonds are called saturated fatty acids (Figure 2.10).

An unsaturated fatty acid

A saturated fatty acid

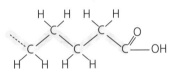

Double bond

Figure 2.10 Unsaturated and saturated fatty acids

Lipids containing unsaturated fatty acids are called unsaturated lipids, and those containing completely saturated fatty acids are called saturated lipids. Animal lipids are often saturated lipids. Plant lipids are often unsaturated. Unsaturated lipids tend to have lower melting points than saturated lipids.

Triglycerides are used as energy storage compounds in plants, animals and fungi. Their insolubility in water helps to make them suitable for this function. They contain more energy per gram than polysaccharides, so can store more energy in less mass. In mammals, stores of triglycerides often build up beneath the skin, in the form of adipose tissue. The cells in adipose tissue contain oil droplets made up of triglycerides.

This tissue also helps to insulate the body against heat loss. It is a relatively low-density tissue, and therefore increases buoyancy. These properties make it especially useful for aquatic mammals that live in cold water, such as whales and seals.

Adipose tissue also forms a protective layer around some of the body organs, for example the kidneys.

In plants, triglycerides often make up a major part of the energy stores in seeds, either in the cotyledons (e.g. in sunflower seeds) or in the endosperm (e.g. in castor beans).

Phospholipids Revised

A phospholipid molecule is like a triglyceride in which one of the fatty acids is replaced by a phosphate group (Figure 2.11).

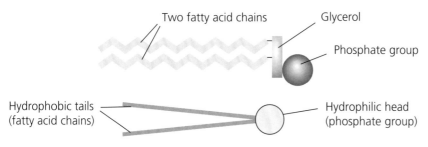

Figure 2.11 Phospholipid molecules

The fatty acid chains have no electrical charge and so are not attracted to the dipoles of water molecules (see p. 23). They are said to be **hydrophobic**.

The phosphate group has an electrical charge and is attracted to water molecules. It is **hydrophilic**.

In water, a group of phospholipid molecules therefore arranges itself into a bilayer, with the hydrophilic heads facing outwards into the water and the hydrophobic tails facing inwards, therefore avoiding contact with water (Figure 2.12).

Figure 2.12 A phospholipid bilayer

This is the basic structure of a cell membrane. The functions of phospholipids in membranes are described on p. 33.

Test for lipids Revised

Mix the substance to be tested with absolute ethanol. Decant the ethanol into water. A milky emulsion indicates the presence of lipid.

Proteins

Proteins are large molecules made of long chains of amino acids.

Amino acids Revised

All amino acids have the same basic structure, with an amine group and a carboxyl group attached to a central carbon atom (Figure 2.13). There are twenty different types of amino acid, which differ in the atoms present in the R group. In the simplest amino acid, glycine, the R group is a single hydrogen atom.

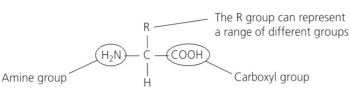

Figure 2.13 An amino acid

Two amino acids can link together by a condensation reaction to form a dipeptide. The bond that links them is called a **peptide bond**, and water is produced in the reaction (Figure 2.14).

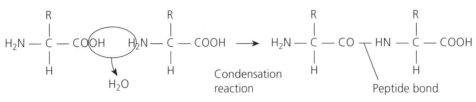

Figure 2.14 Formation of a dipeptide

The dipeptide can be broken down in a hydrolysis reaction, which breaks the peptide bond with the addition of a molecule of water (Figure 2.15).

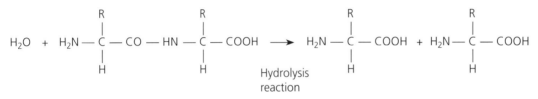

Figure 2.15 Breakdown of a dipeptide

Structure of protein molecules

Revised

Amino acids can be linked together in any order to form a long chain called a **polypeptide**. A polypeptide can form a protein molecule on its own, or it can associate with other polypeptides to form a protein molecule.

The sequence of amino acids in a polypeptide or protein molecule is called its **primary structure** (Figure 2.16). Note that the three letters in each box are the first three letters of the amino acid, for example Val is valine and Leu is leucine.

Val – Leu – Ser – Pro – Ala – Asp – Lys – Thr – Asn – Val – Lys – Ala

Figure 2.16 The primary structure of a small part of a polypeptide

The chain of amino acids often folds or curls up on itself. For example, many polypeptide chains coil into a regular 3D shape called an **alpha helix**. This is held in shape by **hydrogen bonds** between amino acids at different places in the chain. This regular shape is an example of **secondary structure** of a protein (Figure 2.17). Another example is the beta-pleated strand.

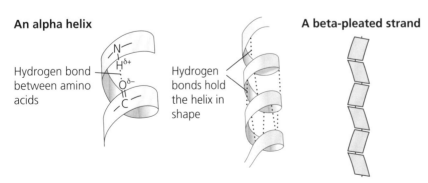

Figure 2.17 Examples of secondary structure

The polypeptide chain can also fold around on itself to form a more complex three-dimensional shape. This is called the **tertiary structure** of the protein. Once again, hydrogen bonds between amino acids at different points in the chain help to hold it in its particular 3-D shape. There are also other bonds involved, including **ionic bonds**, **disulfide bonds** and **hydrophobic interactions** (Figure 2.18).

The globular shape of this polypeptide is an example of tertiary structure

Figure 2.18 Tertiary structure of a protein

Many proteins are made of more than one polypeptide chain. These chains are held together by the same type of bonds as in the tertiary structure (Figure 2.19). The overall structure of the molecule is known as the **quaternary structure** of the protein. The tertiary and quaternary structures of a protein, and therefore its properties, are ultimately determined by its primary structure.

Primary structure is the sequence of amino acids in a polypeptide or protein.

Secondary structure is the first level of folding of the amino acid chain.

Tertiary structure is the second level of folding of the chain.

Quaternary structure is the association of two or more polypeptide chains.

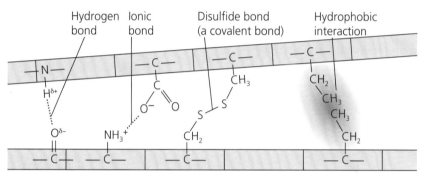

Figure 2.19 Bonds involved in maintaining the secondary, tertiary and quaternary structure of proteins

Globular and fibrous proteins

Revised

Globular proteins have molecules that fold into a roughly spherical three-dimensional shape. Examples include haemoglobin, insulin and enzymes. They are often soluble in water and may be physiologically active — that is, they are involved in metabolic reactions within or outside cells.

Fibrous proteins have molecules that do not curl up into a ball. They have long, thin molecules, which often lie side by side to form fibres. Examples include keratin (in hair) and collagen (in skin and bone). They are not soluble in water and are not generally physiologically active. They often have structural roles.

Haemoglobin — a globular protein

Figure 2.20 shows the structure of a haemoglobin molecule.

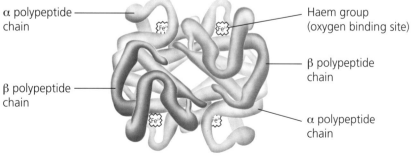

Figure 2.20 The structure of haemoglobin

Relationship between structure and function in haemoglobin

The function of haemoglobin is the transport of oxygen from the lungs to respiring tissues. It is found inside red blood cells.

- **Solubility** The tertiary structure of haemoglobin makes it soluble. The four polypeptide chains are coiled up so that R groups with small charges on them (hydrophilic groups) are on the outside of the molecule. They therefore form hydrogen bonds with water molecules. Hydrophobic R groups are mostly found inside the molecule.

- **Ability to combine with oxygen** The haem group contained within each polypeptide chain enables the haemoglobin molecule to combine with oxygen. Oxygen molecules combine with the iron ion, Fe^{2+}, in the haem group. One oxygen molecule (two oxygen atoms) can combine with each haem group, so one haemoglobin molecule can combine with four oxygen molecules (eight oxygen atoms).

- **Pick-up and release of oxygen** The overall shape of the haemoglobin molecule enables it to pick up oxygen when the oxygen concentration is high, and to release oxygen when the oxygen concentration is low. Small changes in oxygen concentration have a large effect on how much oxygen the haemoglobin molecule can hold. Once one oxygen molecule has combined with one haem group, the whole molecule changes its shape in such a way that it is easier for oxygen to combine with the other three haem groups. (See also information about the oxygen dissociation curve for haemoglobin on pp. 56–57.)

> **Typical mistake**
>
> Although it may sound unlikely, it is common for students to confuse haemoglobin molecules with red blood cells. Remember that each red cell contains millions of haemoglobin molecules.

Collagen — a fibrous protein

Figure 2.21 shows the structure of collagen.

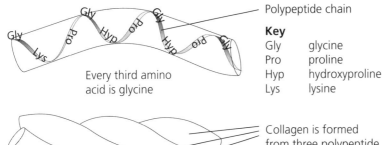

Polypeptide chain

Key
Gly glycine
Pro proline
Hyp hydroxyproline
Lys lysine

Every third amino acid is glycine

Collagen is formed from three polypeptide chains held together by hydrogen bonds

Figure 2.21 The structure of collagen

Relationship between structure and function in collagen

The function of collagen is to provide support and some elasticity in many different animal tissues, such as human skin, bone and tendons.

- **Insolubility** Collagen molecules are very long and are too large to be able to dissolve in water.

- **High tensile strength** Three polypeptide chains wind around one another, held together by hydrogen bonds, to form a three-stranded molecule that can withstand quite high pulling forces without breaking. This structure also allows the molecules to stretch slightly when pulled.

- **Compactness** Every third amino acid in each polypeptide is glycine, whose R group is just a single hydrogen molecule. Their small size allows the three polypeptide chains in a molecule to pack very tightly together.

- **Formation of fibres** There are many lysine molecules in each polypeptide, facing outwards from the three-stranded molecule. This allows covalent bonds to form between the lysine R groups of different collagen molecules, causing them to associate to form fibres.

Test for proteins

Revised

Add biuret solution. A purple colour indicates the presence of protein.

Now test yourself

Tested

> 4 Explain the difference between each of the following:
>   a the secondary structure and the tertiary structure of a protein
>   b a collagen molecule and a collagen fibre
>
> **Answer on p.202**

Revision activities

Make a list of all the polymers described on pp. 16–22. Then construct and complete a table with the following headings:

● Polymer
● Monomers from which it is made
● Bonds linking the monomers
● Functions

Construct and complete a table with the following headings:

● Biological molecule
● How to test for it
● Positive results

Water

About 80% of the body of an organism is water. Water has unusual properties compared with other substances, because of the structure of its molecules. Each water molecule has a small negative charge (δ^-) on the oxygen atom and a small positive charge (δ^+) on each of the hydrogen atoms. This is called a **dipole**.

There is an attraction between the δ^- and δ^+ parts of neighbouring water molecules. This is called a **hydrogen bond** (Figure 2.22).

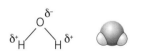
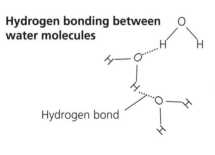

Figure 2.22 Water molecules

Solvent properties of water

Revised

The dipoles on water molecules make water an excellent solvent. For example, if you stir sodium chloride into water, the sodium and chloride ions separate and spread between the water molecules — they dissolve in the water. This happens because the positive charge on each sodium ion is attracted to the small negative charge on the oxygen of the water molecules. Similarly, the negative chloride ions are attracted to the small positive charge on the hydrogens of the water molecules.

Any substance that has fairly small molecules with charges on them, or that can separate into ions, can dissolve in water (Figure 2.23).

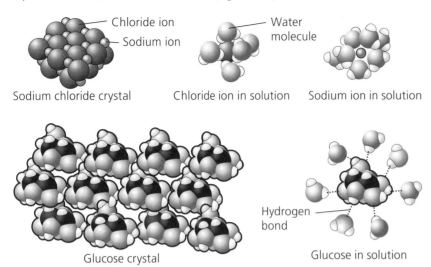

Figure 2.23 Water as a solvent

Because it is a good solvent, water helps to **transport** substances around the bodies of organisms. For example, the blood plasma of mammals is mostly water, and carries many substances in solution, including glucose, oxygen and ions such as sodium. Water also acts as a **medium** in which metabolic reactions can take place, as the reactants are able to dissolve in it.

Thermal properties of water

Revised

- **Water is liquid at normal Earth temperatures**. The hydrogen bonds between water molecules prevent them flying apart from each other at normal temperatures on Earth. Between 0 °C and 100 °C, water is in the liquid state. The water molecules move randomly, forming transitory hydrogen bonds with each other. Other substances whose molecules have a similar structure, such as hydrogen sulfide (H_2S), are gases at these temperatures, because there are no hydrogen bonds to attract their molecules to each other.

- **Water has a high latent heat of evaporation**. When a liquid is heated, its molecules gain kinetic energy, moving faster. Those molecules with the most energy are able to escape from the surface and fly off into the air. A great deal of heat energy has to be added to water molecules before they can do this, because the hydrogen bonds between them have to be broken. When water evaporates, it therefore absorbs a lot of heat from its surroundings. The evaporation of water from the skin of mammals when they sweat therefore has a cooling effect. Transpiration from plant leaves is important in keeping them cool in hot climates.

- **Water has a high specific heat capacity**. Specific heat capacity is the amount of heat energy that has to be added to a given mass of a substance to raise its temperature by 1 °C. Temperature is related to the kinetic energy of the molecules — the higher their kinetic energy, the higher the temperature. A lot of heat energy has to be added to water to raise its temperature, because much of the heat energy is used to break the hydrogen bonds between water molecules, not just to increase their speed of movement. This means that bodies of water, such as oceans or a lake, do not change their temperature as easily as air does. It also means that the bodies of organisms, which contain large amounts of water, do not change temperature easily.

Now test yourself

5 a Explain what a hydrogen bond is.

   b Describe how hydrogen bonds affect the properties of water.

   c State two types of macromolecule that contain hydrogen bonds.

Answer on p.202

Tested

3 Enzymes

Enzyme mode of action

An enzyme is a globular protein that acts as a biological catalyst — that is, it speeds up a metabolic reaction without itself being permanently changed. Some enzymes work inside cells, and are called intracellular enzymes. Some work outside cells (for example, inside the human alimentary canal) and are called extracellular enzymes.

The substance present at the start of an enzyme-catalysed reaction is called the **substrate**. The **product** is the new substance or substances formed.

Active sites
Revised

In one part of the enzyme molecule, there is an area called the **active site**, where the substrate molecule can bind. This produces an enzyme–substrate complex. The 3-D shape of the active site fits the substrate perfectly, so only one type of substrate can bind with the enzyme. The enzyme is therefore **specific** for that substrate. The enzyme can be considered to be a lock, and the substrate the key that fits into the lock. However, in most cases the substrate actually causes a small change in the shape of the active site, allowing the two to fit together. This is called the induced-fit hypothesis.

The R groups of the amino acids at the active site are able to form temporary bonds with the substrate molecule. This pulls the substrate molecule slightly out of shape, causing it to react and form products.

Typical mistake

Remember that it is the enzyme that has an active site, not the substrate.

Typical mistake

Do not say that the enzyme's active site and the substrate have the 'same' shape. Their shapes are complementary.

Activation energy
Revised

Substrates generally need to be supplied with energy to cause them to change into products. The energy required to do this is called **activation energy**. In a laboratory, you might supply energy by heating to cause two substances to react together.

Enzymes are able to make substances react even at low temperatures. They reduce the activation energy needed to make the reaction take place. They do this by distorting the shape of the substrate molecule when it binds at the enzyme's active site.

You can follow what happens over time in a reaction catalysed by an enzyme by:

- measuring the rate of formation of the product
- measuring the rate of disappearance of the substrate

For example, you can measure the rate of formation of oxygen in this reaction:

$$\text{hydrogen peroxide} \xrightarrow{\text{catalase}} \text{oxygen} + \text{water}$$

All biological material contains catalase. You could mash up some potato tuber or celery stalks, mix them with water and filter the mixture to obtain a solution containing catalase. This can then be added to hydrogen peroxide in a test tube. Use relatively small tubes, so that there is not too much gas in the tube above the liquid.

You could measure the rate of oxygen formation by collecting the gas in a gas syringe and recording the volume every minute until the reaction stops (Figure 3.1). Note: don't worry if you don't have gas syringes. You could collect the oxygen in an inverted measuring cylinder over water instead.

You could measure the rate of disappearance of starch in the reaction:

$$\text{starch} \xrightarrow{\text{amylase}} \text{maltose}$$

Add amylase solution to starch suspension in a test tube. Take samples of the reacting mixture at regular time intervals, and test for the presence of starch using iodine in potassium iodide solution. If starch is still present, you will obtain a blue-black colour. If there is no starch present, the iodine solution will remain orange-brown.

To obtain quantitative results, you could use a colorimeter. Put some of the iodine solution into one of the colorimeter tubes, place it in the colorimeter and adjust the dial to give a reading of 0. This is your standard, with no starch. Every minute, take a sample of the liquid from the starch–amylase mixture and add it to a clean colorimeter tube containing iodine solution. Mix thoroughly, then measure the absorbance (Figure 3.2). The darker the blue-black colour, the greater the absorbance, and the greater the concentration of starch.

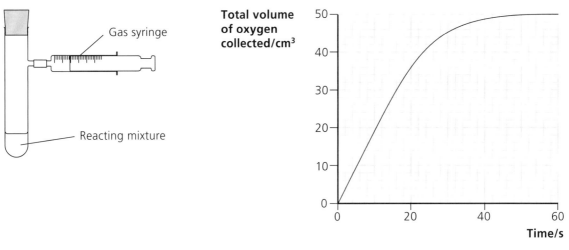

Figure 3.1 Following the time course of the breakdown of hydrogen peroxide by catalase

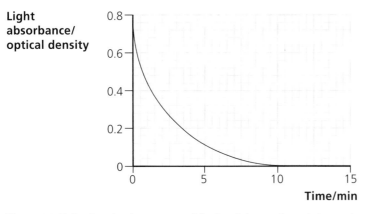

Figure 3.2 Following the time course of the breakdown of starch by amylase

Factors affecting the rate of enzyme-catalysed reactions

When an enzyme solution is added to a solution of its substrate, the random movements of enzyme and substrate molecules cause them to collide with each other.

As time passes, the quantity of substrate decreases, because it is being changed into product. This decrease in substrate concentration means that the frequency of collisions between enzyme and substrate molecules decreases, so the rate of the reaction gradually slows down. The reaction rate is fastest right at the start of the reaction, when substrate concentration is greatest.

When comparing reaction rates of an enzyme in different circumstances, we should therefore try to measure the initial rate of reaction — that is, the rate of reaction close to the start of the reaction.

Temperature Revised

At low temperatures, enzyme and substrate molecules have little kinetic energy. They move slowly, and so collide infrequently. This means that the rate of reaction is low. If the temperature is increased, then the kinetic energy of the molecules increases. Collision frequency therefore increases, causing an increase in the rate of reaction.

Above a certain temperature, however, hydrogen bonds holding the enzyme molecule in shape begin to break. This causes the tertiary structure of the enzyme to change, an effect called **denaturation**. This affects the shape of its active site. It becomes less likely that the substrate molecule will be able to bind with the enzyme, and the rate of reaction slows down.

The temperature at which an enzyme works most rapidly, just below that at which denaturation begins, is called its optimum temperature (Figure 3.3). Enzymes in the human body generally have an optimum temperature of about 37 °C, but enzymes from organisms that have evolved to live in much higher or lower temperatures may have much higher or lower optimum temperatures.

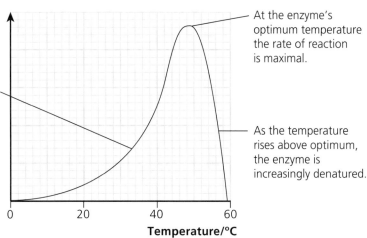

Figure 3.3 How temperature affects the rate of an enzyme-catalysed reaction

How to... Investigate the effect of temperature on enzyme activity

You can use almost any enzyme reaction for this, such as the action of catalase on hydrogen peroxide, as described on p. 26. You could use the same method of collecting the gas that is described there, but here is another possible method.

Set up several small conical flasks containing the same volume of hydrogen peroxide solution of the same concentration. Stand each one in a water bath at a particular temperature. Use at least five different temperatures over a good range — say between 0 °C and 90 °C. (If time allows, set up three sets of tubes at each temperature. You will then be able to calculate the mean result for each temperature, which will give you a more representative finding.)

Take a set of test tubes and add the same volume of catalase solution to each one. Stand these in the same set of water baths.

Leave all the flasks and tubes to come to the correct temperature. Check with a thermometer.

Take the first flask, dry its base and sides and stand it on a sensitive top-pan balance. Pour in the solution containing catalase (see p. 26) that is at the same temperature, and immediately take the balance reading. Record the new balance readings every 30 seconds (or even more frequently if you can manage it) for about 3 minutes. The readings will go down as oxygen is given off. This represents the mass of product formed.

Repeat with the solutions kept at each of the other temperatures.

Work out the initial rate of each reaction, either taken directly from your readings, or by drawing a graph of mass lost (which is the mass of oxygen) against time for each temperature, and then working out the gradient of the graph over the first 30 seconds or 60 seconds of the reaction (Figure 3.4).

Now you can use your results to plot a graph of initial rate of reaction (y-axis) against temperature.

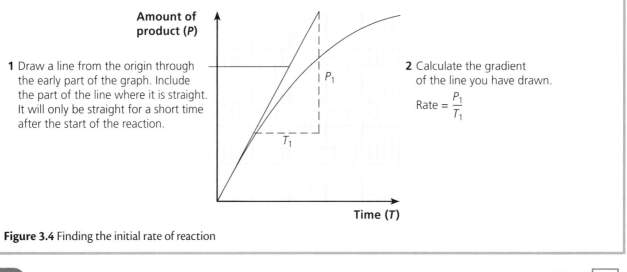

Amount of product (P)

1 Draw a line from the origin through the early part of the graph. Include the part of the line where it is straight. It will only be straight for a short time after the start of the reaction.

P_1

T_1

2 Calculate the gradient of the line you have drawn.

$$\text{Rate} = \frac{P_1}{T_1}$$

Time (T)

Figure 3.4 Finding the initial rate of reaction

pH

Revised

pH affects ionic bonds that hold protein molecules in shape. Because enzymes are proteins, their molecules are affected by changes in pH. Most enzyme molecules only maintain their correct tertiary structure within a very narrow pH range (Figure 3.5), generally around pH 7. Some, however, require a very different pH; one example is the protein-digesting enzyme pepsin found in the human stomach, which has an optimum pH of 2.

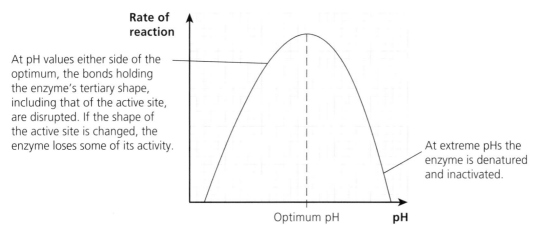

Rate of reaction

At pH values either side of the optimum, the bonds holding the enzyme's tertiary shape, including that of the active site, are disrupted. If the shape of the active site is changed, the enzyme loses some of its activity.

At extreme pHs the enzyme is denatured and inactivated.

Optimum pH   **pH**

Figure 3.5 How pH affects the rate of an enzyme-catalysed reaction

How to... Investigate the effect of pH on enzyme activity

You can adapt the method described in How to... on p. 28 to investigate the effect of pH on the rate of breakdown of hydrogen peroxide by catalase.

Vary pH by using different buffer solutions added to each enzyme solution. (A buffer solution keeps a constant pH, even

if acidic or alkaline products are formed.) Keep temperature, enzyme concentration, substrate concentration and total volume of reactants the same for all the tubes. Record, process and display results as before.

Enzyme concentration
Revised

The greater the concentration of enzyme, the more frequent the collisions between enzyme and substrate, and therefore the faster the rate of the reaction. However, at very high enzyme concentrations, the concentration of substrate may become a limiting factor, so the rate does not continue to increase if the enzyme concentration is increased (Figure 3.6).

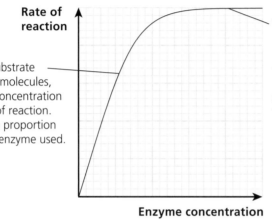

When there are more substrate molecules than enzyme molecules, increasing the enzyme concentration produces a higher rate of reaction. The rate of reaction is in proportion to the concentration of enzyme used.

When there are more enzyme molecules than substrate molecules, increasing the enzyme concentration does not result in a higher rate of reaction as there are no spare substrate molecules for the enzyme to act on.

Figure 3.6 How enzyme concentration affects the rate of an enzyme-catalysed reaction

How to... Investigate the effect of enzyme concentration on rate of reaction

You could use the following method to investigate the effect of enzyme concentration on the rate at which the enzyme catalase converts its substrate, hydrogen peroxide, to water and oxygen.

Prepare a catalase solution as described on p. 26.

Prepare different dilutions of this solution (Table 3.1).

Table 3.1

Volume of initial solution/cm³	Volume of distilled water added/cm³	Relative concentration of catalase (as a percentage of the concentration of the initial solution)
10	0	100
9	1	90
8	2	80

Continue until the final 'solution' prepared is 10 cm³ of distilled water.

Place each solution into a tube fitted with a gas syringe (see p. 26). Use relatively small tubes, so that there is not too much gas in the tube above the liquid, but leave space to add an equal

volume of hydrogen peroxide solution at the next step. Ensure that each tube is labelled with a waterproof marker. If time and materials allow, prepare three sets of these solutions.

Place each tube in a water bath at 30 °C.

Take another set of tubes and add 10 cm³ of hydrogen peroxide solution to each one. The concentration of hydrogen peroxide must be the same in each tube. Stand these tubes in the same water bath.

Leave all the tubes for at least 5 minutes to allow them to come to the correct temperature. When ready, add the contents of one of the hydrogen peroxide tubes to the first enzyme tube. Mix thoroughly. Measure the volume of gas collected in the gas syringe after 2 minutes. If you are using three sets, then repeat using the other two tubes containing the same concentration of enzyme.

Do the same for each of the tubes of enzyme. Record the mean volume of gas produced in 2 minutes for each enzyme concentration and plot a line graph to display your results.

Note: if you find that you get measurable volumes of gas sooner than 2 minutes after mixing the enzyme and substrate, then take your readings earlier. The closer to the start of the reaction you make the measurements, the better.

Substrate concentration

Revised

The greater the concentration of substrate, the more frequent the collisions between enzyme and substrate, and therefore the faster the rate of the reaction. However, at high substrate concentrations, the concentration of enzyme may become a limiting factor, so the rate does not continue to increase if the substrate concentration is increased (Figure 3.7). The maximum rate that is reached is known as V_{max}.

The **Michaelis–Menten constant**, K_m, can be used to compare the affinity of different enzymes for their substrates. K_m is the substrate concentration at which the rate of activity of the enzyme is $\frac{1}{2}V_{max}$. The lower the value of K_m, the greater the affinity of the enzyme for its substrate.

> V_{max} is the maximum rate of activity of the enzyme before substrate concentration becomes limiting.
>
> The **Michaelis–Menten constant**, K_m, is the substrate concentration at which the rate of enzyme activity is $\frac{1}{2}V_{max}$.

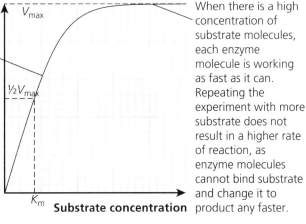

When there is a low concentration of substrate molecules, repeating the experiment with more substrate produces a higher rate of reaction. The rate of reaction is in proportion to the concentration of substrate.

When there is a high concentration of substrate molecules, each enzyme molecule is working as fast as it can. Repeating the experiment with more substrate does not result in a higher rate of reaction, as enzyme molecules cannot bind substrate and change it to product any faster.

Figure 3.7 How substrate concentration affects the rate of an enzyme-catalysed reaction

> ## Now test yourself
>
> 1 Outline how you could compare the affinity of catalase from two different fruits for their substrate.
>
> ### Answer on pp.202–203
>
> Tested

How to... Investigate the effect of substrate concentration on the rate of an enzyme catalysed reaction

You can do this in the same way as described for investigating the effect of enzyme concentration, but this time keep the concentration of catalase the same and vary the concentration of hydrogen peroxide.

Inhibitors

Revised

An inhibitor is a substance that slows down the rate at which an enzyme works (Figure 3.8). Many inhibitors are reversible — that is, they do not attach permanently to the enzyme.

Competitive inhibitors generally have a similar shape to the enzyme's normal substrate. They can fit into the enzyme's active site, preventing the substrate from binding. The greater the proportion of inhibitor to substrate in the mixture, the more likely it is that an inhibitor molecule, and not a substrate molecule, will bump into an active site. The degree to which a competitive inhibitor slows down a reaction is therefore affected by the relative concentrations of the inhibitor and the substrate.

Non-competitive inhibitors do not have the same shape as the substrate, and they do not bind to the active site. They bind to a different part of the enzyme. This changes the enzyme's shape, including the shape of the active site, so the substrate can no longer bind with it. Even if you add more substrate, it still will not be able to bind, so the degree to which a non-competitive inhibitor slows down a reaction is *not* affected by the relative concentrations of the inhibitor and the substrate.

Enzyme acting on its normal substrate

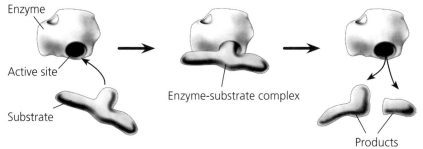

Enzyme

Active site

Substrate

Enzyme-substrate complex

Products

Enzyme in the presence of a competitive inhibitor

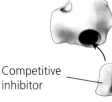

Competitive inhibitor

With the competitive inhibitor bound at the active site, the normal substrate cannot bind.

Enzyme in the presence of a non-competitive inhibitor

Non-competitive inhibitor

With the non-competitive inhibitor bound to the enzyme, the active site is changed and the normal substrate cannot bind.

Figure 3.8 Competitive and non-competitive enzyme inhibitors

Now test yourself

2 Temperature, pH, enzyme concentration, substrate concentration and inhibitors can all affect the rate of enzyme activity.

   a Which of these have their effect by changing the frequency of collisions between the substrate and the enzyme's active site?

   b Which have their effect by changing the shape of the active site?

Answer on p.203

Tested

Immobilising enzymes

Revised

Enzymes are used as catalysts in many industrial processes. The enzymes are often immobilised by fixing them onto or inside a gel or other substance, which prevents the enzymes from dissolving in the reacting mixture. This means that the product is not contaminated with dissolved enzyme, and the enzymes can be repeatedly reused. Immobilisation also enables enzymes to work over a wider range of temperature and pH than when they are in free solution, probably because the trapped enzyme molecules cannot easily change shape and become denatured. One way of immobilising enzymes is to trap them inside little balls of calcium alginate.

How to... Investigate the effect of immobilisation on the rate of an enzyme-catalysed reaction

You can immobilise enzymes inside calcium alginate balls. First, make up a solution of 1.5% calcium chloride, and another of 2% sodium alginate. Mix the chosen enzyme solution into the sodium alginate solution. Using a pipette, draw up a small volume of the mixture, and drop it into the calcium chloride solution. The calcium chloride and sodium alginate will react to form little beads of jelly-like calcium alginate, with the enzyme trapped in them. You can now compare the rate of activity of the enzyme when trapped in the beads, and when free in solution.

Now test yourself

Tested

3 List the reasons why immobilised enzymes, rather than enzymes in solution, are often used in industrial processes.

Answer on p.203

4 Cell membranes and transport

Fluid mosaic membranes

Every cell is surrounded by a cell membrane. There are also many membranes within cells. The membrane around the outside of a cell is called the cell surface membrane.

Structure of a cell membrane

Revised

A cell membrane consists of a double layer of **phospholipid** molecules (p. 19). This structure arises because in water a group of phospholipid molecules arranges itself into a **bilayer**, with the hydrophilic heads facing outwards into the water and the hydrophobic tails facing inwards, therefore avoiding contact with water.

This is the basic structure of a cell membrane. There are also **cholesterol** molecules among the phospholipids. **Protein** molecules float in the phospholipid bilayer. Many of the phospholipids and proteins have short chains of carbohydrates attached to them, on the outer surface of the membrane. They are known as **glycolipids** and **glycoproteins**. There are also other types of glycolipid with no phosphate groups.

This is called the **fluid mosaic model** of membrane structure (Figure 4.1):
- 'Fluid' because the molecules within the membrane can move around within their own layers.
- 'Mosaic' because the protein molecules are dotted around within the membrane.
- 'Model' because no-one has ever seen a membrane looking like the diagram — the molecules are too small to see even with the most powerful microscope. The structure has been worked out because it explains the behaviour of membranes that has been discovered through experiment.

The roles of the components of cell membranes are outlined in Table 4.1.

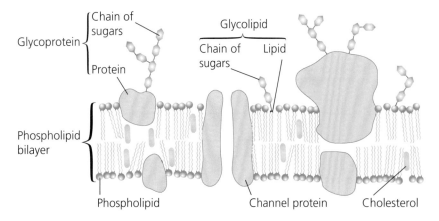

Figure 4.1 The fluid mosaic model of membrane structure

Table 4.1 Cell membrane components and their roles

Component	Roles
Phospholipids	Form the fluid bilayer that is the fundamental structure of the membrane.
	Prevent hydrophilic substances — such as ions and some molecules — from passing through.
Cholesterol	Helps to keep the cell membrane fluid.
Proteins and glycoproteins	Provide channels that allow hydrophilic substances to pass through the membrane; these channels can be opened or closed to control the substances' movement.
	Actively transport substances through the membrane against their concentration gradient, using energy derived from ATP.
	Act as receptor molecules for substances such as hormones, which bind with them; this can then affect the activity of the cell.
	Cell recognition — cells from a particular individual or a particular tissue have their own set of proteins and glycoproteins on their outer surfaces.
Glycolipids	Cell recognition and adhesion to neighbouring cells to form tissues.

Cell signalling

Revised

Cells can communicate with one another using molecules that interact with cell membranes. Cell surface membranes contain receptor molecules, into which signalling molecules can fit. Like enzymes, these receptors are specific, only accepting one type of signalling molecule. This means that the signalling molecule can only affect cells that have its receptor in their cell surface membranes. These are its target cells.

For example, the hormone insulin is a protein that fits into receptors in the cell surface membrane of liver cells. When insulin is bound to the receptor, this brings about changes in the cell that result in an increase of transporter proteins for glucose in the cell surface membrane, causing the cell to take up glucose.

Movement of substances into and out of cells

Passive transport through cell membranes

Revised

Molecules and ions are in constant motion. In gases and liquids they move freely. As a result of their random motion, each type of molecule or ion tends to spread out evenly within the space available. This is **diffusion**. Diffusion results in the net movement of ions and molecules from a high concentration to a low concentration.

> **Diffusion** is the net movement of molecules or ions down a concentration gradient, as a result of the random movement of particles.

Diffusion across a cell membrane

Some molecules and ions are able to pass through cell membranes. The membrane is permeable to these substances. However, some substances cannot pass through cell membranes, so the membranes are said to be **partially permeable**.

For example, oxygen is often at a higher concentration outside a cell than inside, because the oxygen inside the cell is being used up in respiration. The random motion of oxygen molecules inside and outside the cell means that some of them 'hit' the cell surface membrane. More of them hit the membrane on the outside than the inside, because there are more of them outside. Oxygen molecules are small and do not carry an electrical charge, so they are able to pass freely through the phospholipid bilayer. Oxygen therefore diffuses from outside the cell, through the membrane, to the inside of the cell, down its concentration gradient.

This is passive transport, because the cell does not do anything to cause the oxygen to move across the cell membrane.

How to... Investigate the effect of surface area on diffusion rate

In general, large objects have a smaller surface area:volume ratio than small objects. You can use blocks of agar jelly to investigate how the rate of diffusion is affected by surface area:volume ratio. The agar jelly can be made up using a little universal indicator solution. If the water used to dissolve the agar is slightly acidic, then the jelly will be red. Decide on the shapes and sizes of jelly to cut, thinking about keeping the volume of each piece constant, but varying the surface area. Immerse your pieces of jelly in a dilute alkali (e.g. sodium hydrogencarbonate solution). As the alkali diffuses into the jelly, the indicator will change colour. You can time how long it takes for the whole piece of jelly to change colour.

Facilitated diffusion

Ions or electrically charged molecules are not able to diffuse through the phospholipid bilayer because they are repelled from the hydrophobic tails. Large molecules are also unable to move through the phospholipid bilayer freely. However, the cell membrane contains special protein molecules that provide hydrophilic passageways through which these ions and molecules can pass. They are called **channel proteins**. Different channel proteins allow the passage of different types of molecules and ions. Diffusion through these channel proteins is called **facilitated diffusion**. Like 'ordinary' diffusion, it is entirely passive (Figure 4.2).

> **Facilitated diffusion** is the net movement of molecules or ions through channel proteins in a cell membrane, down a concentration gradient, as a result of the random movement of particles.

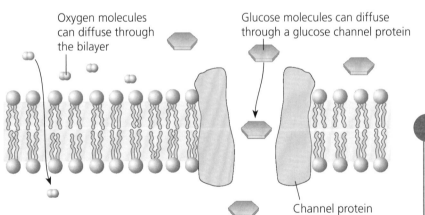

Figure 4.2 Diffusion across a cell membrane

Oxygen molecules can diffuse through the bilayer

Glucose molecules can diffuse through a glucose channel protein

Channel protein

Now test yourself

1 Explain the difference between diffusion and facilitated diffusion.

Answer on p.203

Tested

Osmosis

Water molecules are small. They carry tiny electrical charges (dipoles) but their small size means that they are still able to move quite freely through the phospholipid bilayer of most cell membranes. Water molecules therefore tend to diffuse down their concentration gradient across cell membranes.

Cell membranes always have a watery solution on each side. These solutions may have different concentrations of solutes.

The greater the concentration of solute, the less water is present. The water molecules in a concentrated solution are also less free to move, because they are attracted to the solute molecules (see Figure 4.3). A concentrated solution is therefore said to have a **low water potential**.

In a dilute solution, there are more water molecules and they can move more freely. This solution has a **high water potential**.

Imagine a cell membrane with a dilute solution on one side and a concentrated solution on the other side. The solute has molecules that are too large to get through the membrane — only the water molecules can get through. The membrane is said to be **partially permeable**.

Water molecules in the dilute solution are moving more freely and therefore hit the membrane more often than water molecules in the concentrated solution. More water molecules therefore diffuse across the membrane from the dilute to the concentrated solution than in the other direction. The net movement of water molecules is from a high water potential to a low water potential, down a water potential gradient. This is **osmosis** (Figure 4.3).

Osmosis is the diffusion of water molecules from a dilute solution to a concentrated solution through a partially permeable membrane, down a water potential gradient.

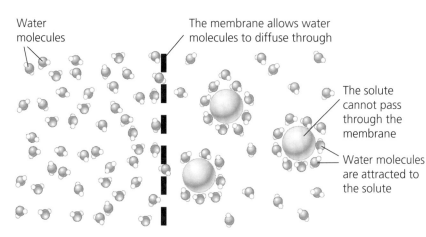

Water molecules

The membrane allows water molecules to diffuse through

The solute cannot pass through the membrane

Water molecules are attracted to the solute

Figure 4.3 Osmosis across a cell membrane

How to... Use Visking tubing to investigate osmosis

You can use an artificial partially permeable membrane called Visking tubing to investigate osmosis. The apparatus in Figure 4.4 could be used to investigate how temperature or different concentrations of solutions (differences in water potential) affect the rate of osmosis.

Your results will be recorded as the level reached by the liquid in the narrow glass tube at regular time intervals

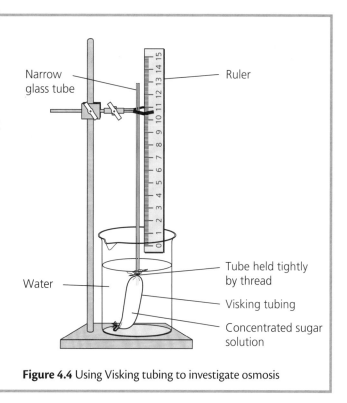

Narrow glass tube

Ruler

Water

Tube held tightly by thread

Visking tubing

Concentrated sugar solution

Figure 4.4 Using Visking tubing to investigate osmosis

Water potential is measured in pressure units, kilopascals (kPa). Pure water has a water potential of 0 kPa. Solutions have negative water potentials. For example, a dilute sucrose solution might have a water potential of −250 kPa. A concentrated sucrose solution might have a water potential of −4000 kPa. The more negative the number, the lower the water potential. Water moves by osmosis down a water potential gradient, from a high (less negative) water potential to a low (more negative) water potential (Figure 4.5).

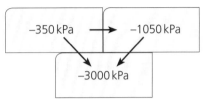

Net water movement occurs in the direction of the arrows, from cells with high water potential to cells with low water potential

Figure 4.5 Water movement between cells

Effects of osmosis on animal and plant cells

An animal cell placed in pure water takes up water by osmosis, as water molecules diffuse down the water potential gradient into the cell. The volume of the cell increases, and the cell may burst. Animals and protoctists (single-celled organisms with animal-type cells) that live in fresh water have some way of removing excess water from their cells, so that the cells do not swell and burst.

An animal cell in a concentrated solution loses water by osmosis, and will shrink.

A plant cell in pure water also takes up water by osmosis. However, it has a strong cell wall outside its cell surface membrane, which prevents the cell from bursting. Instead, the cell simply becomes swollen and is said to be turgid. Turgidity helps plant tissues to remain in shape. For example, a plant leaf remains in shape because of the turgidity of its cells.

A plant cell in a concentrated solution loses water by osmosis, so that the volume of the cytoplasm and vacuole decrease. The cell loses its turgidity. Leaves in which the cells lose their turgidity are no longer supported, and they wilt.

If a great deal of water is lost from a plant cell, then the cell contents shrink so much that the cell surface membrane pulls away from the cell wall. This is called plasmolysis (Figure 4.6). The membrane often remains attached at points where plasmodesmata link to the next cell.

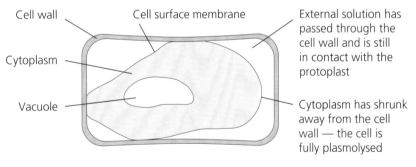

Cell wall

Cell surface membrane

External solution has passed through the cell wall and is still in contact with the protoplast

Cytoplasm

Vacuole

Cytoplasm has shrunk away from the cell wall — the cell is fully plasmolysed

Figure 4.6 A plasmolysed plant cell

Investigate the effect on plant cells of immersion in solutions of different water potentials

There are several different ways in which this investigation could be done. They include the following:

- Cut cylinders or discs or strips of a solid and uniform plant tissue, such as a potato tuber, then measure either their lengths or masses before immersing them in the solutions. Leave them long enough for them to come to equilibrium, and then measure the length or mass of each piece again, and calculate the percentage change in the measurement. Percentage change can then be plotted against the concentration of the solution.
- Cut small pieces of single-cell-thick plant tissue, for example onion epidermis. Mount them in a drop of sugar solution on microscope slides, and count the percentage of cells plasmolysed, or score each cell you see according to how plasmolysed it is.

For the plasmolysis investigation, you could use this method:

- Peel off one of the thick layers from an onion bulb. Cut six approximately 1 cm² pieces from it. Put each piece into a different liquid. One should be distilled water, then a range of sucrose solutions from about 0.1 mol dm⁻³ up to about 1.0 mol dm⁻³.
- Take six clean microscope slides and label each with the concentration of solution you are going to place on it. Put a drop of the relevant solution on each one.

- Peel off the inner epidermis from one of the 1 cm² pieces of onion, and place it carefully onto the drop on a slide of the same concentration solution in which it has been immersed. Take care not to let it fold over. Press it gently into the liquid using a section lifter or other blunt tool. Carefully place a coverslip over it, taking care not to trap air bubbles.
- Repeat with each of the other drops of liquid. Leave all of the slides for at least 5 minutes to give any water movements by osmosis time to take place and equilibrium to be reached.
- Observe each slide under the microscope. Count the total number of cells in the field of view and record this. Then count the number that are plasmolysed and record this. Then move to another area of the slide and repeat until you have counted at least 50 cells. Repeat for each slide.
- Calculate the percentage of cells that have plasmolysed in each solution. Add this to your results chart.
- Plot percentage of cells that have plasmolysed (*y*-axis) against the concentration of the solution (*x*-axis). Join the points with either straight lines drawn between points, or a best-fit curve.
- The point at which the line crosses the 50% plasmolysis level tells you the concentration of the solution at which, on average, cells were just beginning to plasmolyse. At this value, the concentration of the solution inside the onion cells was, on average, the same as the concentration of the sucrose solution.

Active transport across cell membranes

Revised

Cells are able to make some substances move across their membranes *up* their concentration gradients. For example, if there are more potassium ions inside the cell than outside the cell, the potassium ions would diffuse out of the cell. However, the cell may require potassium ions. It may therefore use a process called **active transport** to move potassium ions from outside the cell to inside the cell, against the direction in which they would naturally diffuse.

This is done using carrier (transporter) proteins in the cell membrane. These use energy from the breakdown of ATP to move the ions into the cell. The carrier proteins are ATPases (Figure 4.7).

$$ATP \xrightarrow{ATPase} ADP + phosphate + energy$$

Each carrier protein is specific to just one type of ion or molecule. Cells contain many different carrier proteins in their membranes.

> **Active transport** is the movement of molecules or ions across a cell membrane against their concentration gradient, using energy from respiration.

> **Now test yourself**
>
> 2 Explain the difference between facilitated diffusion and active transport.
>
> **Answer on p.203**
>
> Tested

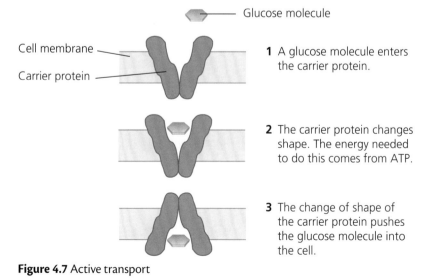

Glucose molecule

Cell membrane

Carrier protein

1 A glucose molecule enters the carrier protein.

2 The carrier protein changes shape. The energy needed to do this comes from ATP.

3 The change of shape of the carrier protein pushes the glucose molecule into the cell.

Figure 4.7 Active transport

Endocytosis and exocytosis

Cells can move substances into and out of the cell without the substances having to pass through the cell membrane.

In **endocytosis** the cell puts out extensions around the object to be engulfed. The membrane fuses together around the object, forming a vesicle.

In **exocytosis** the object is surrounded by a membrane inside the cell to form a vesicle. The vesicle is then moved to the cell membrane. The membrane of the vesicle fuses with the cell membrane, expelling its contents outside the cell (Figure 4.8).

Exocytosis

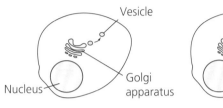 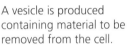   

A vesicle is produced containing material to be removed from the cell.

The vesicle moves to the cell surface membrane.

The membrane of the vesicle and the cell surface membrane join and fuse.

The contents of the vesicle are released.

Endocytosis

The cell surface membrane grows out.

The object or solution is surrounded.

The membrane breaks and rejoins, enclosing the object.

The vesicle moves inwards and its contents are absorbed.

Figure 4.8 Endocytosis and exocytosis

Table 4.2 summarises the methods of movement across membranes.

Table 4.2 Movement across cell membranes

Feature	Passive movement			Active movement	
	Diffusion	Osmosis	Facilitated diffusion	Active transport	Endocytosis and exocytosis
Requires energy input from the cell	No	No	No	Yes	Yes
Involves the movement of individual ions or molecules	Yes	Yes	Yes	Yes	No
Movement is through protein channels or protein carriers in the membrane	No	No	Yes	Yes	No
Examples of substances that move	Oxygen, carbon dioxide	Water	Ions (e.g. K^+, Cl^-) and molecules (e.g. glucose)	Mostly ions (e.g. K^+, Cl^-)	Droplets of liquid; bacteria; export proteins

5 The mitotic cell cycle

Division of cells

Chromosomes

A chromosome is a molecule of **DNA**, associated with proteins called **histones**. The DNA and histones are tightly coiled, so that the long DNA molecule packs into a very small space.

Before cell division, each molecule of DNA is duplicated. The two identical copies of tightly coiled DNA are called **chromatids**, and they remain attached to one another at a point called a **centromere** (Figure 5.1).

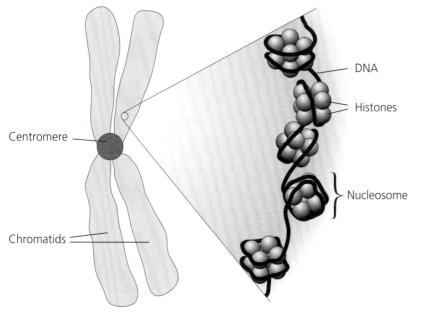

Centromere

Chromatids

DNA

Histones

Nucleosome

Figure 5.1 Chromosome structure

The DNA in a chromosome contains coded instructions for making proteins in the cell (see p. 44), called **genes**. When DNA is duplicated, it cannot be copied right to the very end of the molecule. To avoid loss of genes, the ends of each chromosome consists of **telomeres**, which are regions of non-coding DNA. Each time the DNA is copied, a little of each telomere is lost, so it gradually becomes shorter.

Mitosis

A multicellular organism begins as a single cell. That cell divides repeatedly to produce all the cells in the adult organism.

The type of cell division involved in growth is called **mitosis**. Mitosis is also used to produce new cells to replace ones that have been damaged — that is, to repair tissues. Mitosis is also involved in asexual reproduction, in which a single parent gives rise to genetically identical offspring.

Strictly speaking, mitosis is division of the nucleus of the cell. After this, the cell itself usually divides as well. This is called **cytokinesis**.

The cell cycle Revised

During growth of an organism many of its cells go through a continuous cycle of growth and mitotic division called the **cell cycle** (Figure 5.2).

The two G phases and S phase make up interphase.

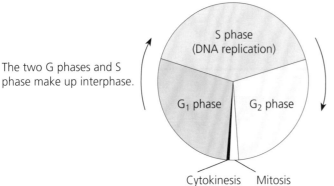

Figure 5.2 The cell cycle

Typical mistake

Students often state that interphase or cytokinesis are stages in mitosis. You can see from Figure 5.2 that this is not correct.

For most of the cell cycle, the cell continues with its normal activities. It also grows, as the result of the production of new molecules of proteins and other substances, which increase the quantity of cytoplasm in the cell.

DNA replication takes place during interphase, so that there are two identical copies of each DNA molecule in the nucleus. (DNA replication is described on p. 43–44.) Each original chromosome is made up of one DNA molecule, so after replication is complete each chromosome is made of two identical DNA molecules.

During mitosis, the two chromatids split apart and are moved to opposite ends of the cell. A new nuclear envelope then forms around each group. These two nuclei each contain a complete set of DNA molecules identical to those in the original (parent) cell. Mitosis produces two genetically identical nuclei from one parent nucleus (see Figure 5.3).

After mitosis is complete, the cell usually divides into two, with one of the new nuclei in each of the two new cells. The two daughter cells are genetically identical to each other and their parent cell.

Control of the cell cycle

Each cell contains genes that help to control when it divides. It is important that cells divide by mitosis only when they are required to do so. This usually involves signals from neighbouring cells, to which the cell responds by either dividing or not dividing. If this control goes wrong, then cells may not divide when they should (so growth does not take place, or wounds do not heal) or they may divide when they should not (so that a tumour may form).

Mitosis

Prophase
- The chromosomes condense
- The centrioles duplicate
- The centriole pairs move towards each pole
- The spindle begins to form

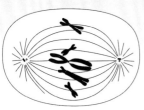

Metaphase
- The nuclear envelope disappears
- The centriole pairs are at the poles
- The spindle is completely formed
- The chromosomes continue to condense
- The spindle fibres attach to the centromeres of the chromosomes
- The spindle fibres pull on the centromeres, arranging them on the equator

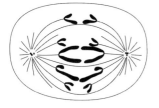

Anaphase
- The links between sister chromatids break
- The centromeres of sister chromatids move apart, pulled by the spindle fibres

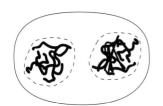

Telophase
- Sister chromatids (now effectively separate chromosomes) reach opposite poles
- The chromosomes decondense
- Nuclear envelopes begin to form around the chromosomes at each pole
- The spindle disappears

Cytokinesis

- The cell divides into two cells, either by infolding of the cell surface membrane in animal cells, or by the formation of a new cell wall and cell surface membrane in plants

Figure 5.3 Mitosis and cytokinesis in an animal cell

Stem cells
Revised

Animals begin their lives as a zygote, when a sperm and egg fuse together at fertilisation. The zygote divides to form two, then four, then eight cells, eventually forming a multicellular embryo. Each of the cells in this embryo is potentially capable of dividing to form any of the many different types of specialised cell in the animal's body.

Cells that are able to divide to form specialised cells are called **stem cells**. In an adult organism, most cells have already become specialised (they have differentiated) and are not able to divide. However, most tissues contain some stem cells that retain the ability to divide, and these are able to form new cells for growth and tissue repair. In a young embryo, all cells are stem cells, and they can form all types of specialised cell. They are said to be totipotent. In an adult, most stem cells are only able to form a limited number of types of specialised cell. For example, stem cells in the bone marrow are able to divide to form different types of blood cell, but they cannot form nerve cells or muscle cells.

6 Nucleic acids and protein synthesis

DNA and RNA

────────────────────────────────── Revised ☐

DNA and RNA are **polynucleotides**. Polynucleotides are substances whose molecules are made of long chains of nucleotides linked together. A nucleotide (Figure 6.1) is made up of:

- a 5-carbon sugar (**deoxyribose** in DNA; **ribose** in RNA)
- a phosphate group
- a nitrogen-containing base (**adenine**, **guanine**, **cytosine** or **thymine** in DNA; adenine, guanine, cytosine or **uracil** in RNA)

The bases are usually referred to by their first letters, A, G, C, T and U.

A and G are **purine** bases, made up of two carbon–nitrogen rings. C, T and U are **pyrimidine** bases, made up of one carbon–nitrogen ring.

Typical mistake

Take care with the spelling of adenine and thymine – do not confuse them with adenosine or thiamine.

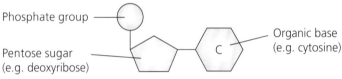

Figure 6.1 labels:
Phosphate group
Pentose sugar (e.g. deoxyribose)
Organic base (e.g. cytosine)
C

Figure 6.1 A nucleotide

Nucleotides can link together by the formation of covalent bonds between the phosphate group of one and the sugar of another (Figure 6.2). This takes place through a condensation reaction.

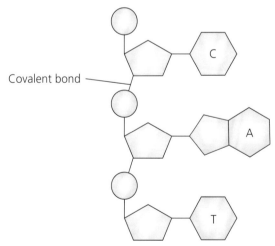

Covalent bond
C
A
T

Figure 6.2 Part of a polynucleotide

An RNA molecule is usually made up of a single strand, although this may be folded up on itself. A DNA molecule is made up of two strands, held together by **hydrogen bonds** between the bases on the two strands. The strands run in opposite directions, i.e. they are anti-parallel.

Hydrogen bonding only occurs between A and T and between C and G. This is called **complementary base pairing** (Figure 6.3).

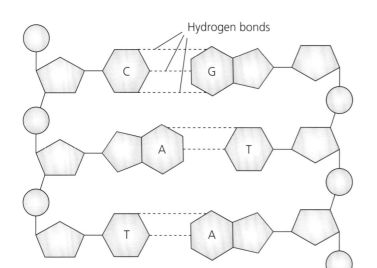

Figure 6.3 Part of a double-stranded DNA molecule

The two strands of nucleotides twist round each other to produce a double helix.

The structure of ATP

Adenosine triphosphate, ATP, is a phosphorylated nucleotide, with a structure similar to an RNA nucleotide (Figure 6.4).

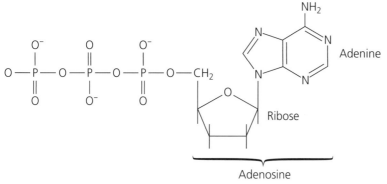

Figure 6.4 An ATP molecule

Now test yourself

1 Use Figure 6.3 to explain why A can only bond with T, not with another A or with C or G.

Answer on p.203

Tested

Expert tip

Do not confuse bases (or nucleotides) with amino acids.

DNA replication

Revised

New DNA molecules need to be made before a cell can divide. The two daughter cells must each receive a complete set of DNA. The base sequences on the new DNA molecules must be identical with those on the original set. DNA replication takes place in the nucleus, during interphase.

● Hydrogen bonds between the bases along part of the two strands are broken. This 'unzips' part of the molecule, separating the two strands.
● Nucleotides that are present in solution in the nucleus are moving randomly around. By chance, a free nucleotide will bump into a newly exposed one with which it can form hydrogen bonds. Free nucleotides therefore pair up with the nucleotides on each of the DNA strands, always A with T and C with G (Figure 6.5).
● DNA polymerase links together the phosphate and deoxyribose groups of adjacent nucleotides.

This is called **semi-conservative replication**, because each new DNA molecule has one old strand and one new one.

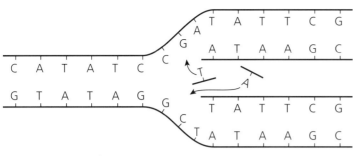

Figure 6.5 DNA replication

The genetic code

The sequence of bases in a DNA molecule is a code that determines the sequence in which amino acids are linked together when making a protein molecule. A sequence of DNA nucleotides that codes for one polypeptide, or for one protein, is known as a **gene**.

As we have seen, the sequence of amino acids in a protein — its primary structure — determines its three-dimensional shape and therefore its properties and functions. For example, the primary structure of an enzyme determines the shape of its active site, and therefore the substrate with which it can bind.

A series of three bases in a DNA molecule, called a base **triplet**, codes for one amino acid. The DNA strand that is used in protein synthesis is called the **template strand**. For example, Figure 6.6 shows the sequence of amino acids coded for by the template strand of a particular length of DNA.

> A **gene** is a sequence of nucleotides that forms part of a DNA molecule, and that codes for a polypeptide or protein.

bases in DNA	T	A	C	C	T	G	C	A	A	C	T	T

amino acid in polypeptide	methionine	aspartate	valine	glutamate

Figure 6.6 A DNA strand and the amino acids it codes for

There are twenty amino acids. Because there are four bases, there are $4^3 = 64$ different possible combinations of bases in a triplet. Some amino acids therefore are coded for by more than one triplet. For example, the triplets AAA and AAG both code for the amino acid phenylalanine. The code is therefore said to be **degenerate**.

Protein synthesis

Revised

Proteins are made on the ribosomes in the cytoplasm, by linking together amino acids through peptide bonds. The sequence in which the amino acids are linked is determined by the sequence of bases on a length of DNA in the nucleus.

Transcription

The first step in protein synthesis is for the sequence of bases on the template strand of the DNA to be used to construct a strand of messenger RNA (**mRNA**) with a complementary sequence of bases. This is called **transcription**.

In the nucleus, the double helix of the DNA is unzipped, exposing the bases on each strand. There are four types of free RNA nucleotide in the nucleus, with the bases A, C, G and U. The RNA nucleotides form hydrogen bonds with the exposed bases on the template strand of the DNA. They pair up as shown in Table 6.1.

> **Typical mistake**
>
> Students sometimes confuse transcription with DNA replication. Read the question carefully and make sure you are writing about the correct process.

Table 6.1 DNA – RNA base pairing

Base on DNA strand	Base on RNA strand
A	U
C	G
G	C
T	A

As the RNA nucleotides slot into place next to their complementary bases on the DNA, the enzyme **RNA polymerase** links them together (through their sugar and phosphate groups) to form a long chain of RNA nucleotides. This is an mRNA molecule.

The mRNA molecule contains a complementary copy of the base sequence on the template strand of part of a DNA molecule. Each triplet on the DNA is represented by a complementary group of three bases on the mRNA, called a **codon** (Figure 6.7).

1 Part of a molecule of DNA

2 The hydrogen bonds between bases are broken, exposing the bases

3 Free RNA nucleotides in the nucleus form new hydrogen bonds with the exposed bases on the template strand

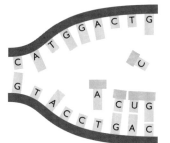

4 The RNA nucleotides are linked together to form an mRNA molecule

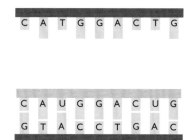

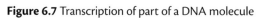

Figure 6.7 Transcription of part of a DNA molecule

Now test yourself

2 How many amino acids are coded for by the length of DNA in Figure 6.7?

Answer on p.203

Tested

Translation

The mRNA molecule breaks away from the DNA, and moves out of the nucleus into the cytoplasm. It becomes attached to a **ribosome**. Two codons fit into a groove in the ribosome. The first codon is generally AUG, which is known as a **start codon**. It codes for the amino acid methionine.

In the cytoplasm, 20 different types of amino acid are present. There are also many different types of transfer RNA (**tRNA**) molecule. Each tRNA molecule is made up of a single strand of RNA nucleotides, twisted round on itself to form a clover-leaf shape. There is a group of three exposed bases, called an **anticodon**. There is also a position at which a particular amino acid can be loaded by a specific enzyme (Figure 6.8).

The amino acid that can be loaded onto the tRNA is determined by the base sequence of its anticodon. For example, a tRNA whose anticodon is UAC will be loaded with the amino acid methionine.

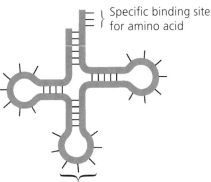

Specific binding site
for amino acid

Three bases forming the anticodon

Figure 6.8 A tRNA molecule

Now test yourself

3   What will the tRNA anticodons be,
    to fit against the mRNA shown in
    Figure 6.7?

Answer on p.203

Tested

The amino acids carried by the two adjacent tRNAs are then linked by a peptide
bond.

The mRNA is then moved along one place in the ribosome, and a third tRNA
slots into place against the next mRNA codon. A third amino acid is added to
the chain.

This continues until a **stop codon** is reached on the mRNA. This is a codon that
does not code for an amino acid, such as UGA. The polypeptide (long chain of
amino acids) that has been formed breaks away.

This process of building a chain of amino acids following the code on an mRNA
molecule is called **translation** (Figure 6.9).

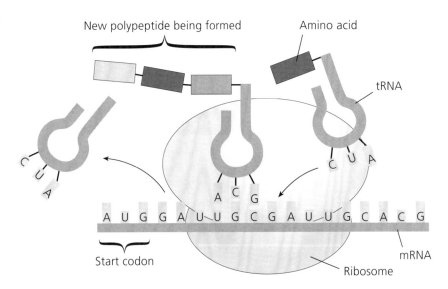

New polypeptide being formed   Amino acid

tRNA

Start codon

mRNA

Ribosome

Figure 6.9 Translation

Mutation

Revised

A mutation is a random, unpredictable change in the DNA in a cell. It may be:

● a change in the sequence of bases in one part of a DNA molecule
● an addition of extra DNA to a chromosome or a loss of DNA from it
● a change in the total number of chromosomes in a cell

Mutations are most likely to occur during DNA replication, for example when
a 'wrong' base might slot into position in the new strand being built. Enzymes
repair almost all these mistakes immediately, but some can persist.

A change in the sequence of bases in DNA can result in a change in the sequence of amino acids in a protein. (Note that this does not always happen, because there is more than one triplet that codes for each amino acid, so a change in a triplet may not change the amino acid that is coded for.) This in turn may result in a change in the 3-D structure of the protein and therefore the way that it behaves.

> A **gene mutation** is a change in the sequence of nucleotides, which may result in an altered polypeptide.

Sickle cell anaemia

An example of a mutation is a change in the gene that codes for one of the polypeptides in a haemoglobin molecule. In the genetic disease **sickle cell anaemia**, the gene that codes for the β polypeptide has the base T where it should have the base A. This means that one triplet is different, so a different amino acid is used when the polypeptide chain is constructed on a ribosome.

These two different forms of the gene are called **alleles**. The normal allele is **HbS**, and the sickle cell allele is **HbA**.

Normal allele

DNA base sequence:

GTG CAC CTG ACT CCT **GAG** GAG AAG TCT

Amino acid sequence:

Val-His-Leu-Thr-Pro-**Glu**-Glu-Lys-Ser-

Abnormal allele

DNA base sequence:

GTG CAC CTG ACT CCT **GTG** GAG AAG TCT

Amino acid sequence:

Val-His-Leu-Thr-Pro-**Val**-Glu-Lys-Ser-

The abnormal β polypeptide has the amino acid valine where it should have the amino acid glutamic acid. These amino acids are on the outside of the haemoglobin molecule when it takes up its tertiary and quaternary shapes.

Glutamic acid is a hydrophilic amino acid. It interacts with water molecules, helping to make the haemoglobin molecule soluble.

Valine is a hydrophobic amino acid. It does not interact with water molecules, making the haemoglobin molecule less soluble.

When the abnormal haemoglobin is in an area of low oxygen concentration, the haemoglobin molecules stick to one another, forming a big chain of molecules that is not soluble and therefore forms long fibres. This pulls the red blood cells (inside which haemoglobin is found) out of shape, making them sickle-shaped instead of round. They are no longer able to move easily through the blood system and may get stuck in capillaries. This is very painful and can be fatal.

7 Transport in plants

Structure of transport tissues

Plants can be very large, but they have a branching shape that helps to keep the surface area-to-volume ratio fairly large. Their energy needs are generally small compared with those of animals, so respiration does not take place so quickly. They can therefore rely on diffusion to supply their cells with oxygen and to remove carbon dioxide. Their leaves are very thin and have a large surface area inside them in contact with the air spaces. This means that diffusion is sufficient to supply the mesophyll cells with carbon dioxide for photosynthesis, and to remove oxygen.

Plant transport systems, therefore, do not transport gases. Plants have two transport systems:

- **xylem**, which transports water and inorganic ions from the roots to all other parts of the plant
- **phloem**, which transports substances made in the plant, such as sucrose and amino acids, to all parts of the plant

How to... Draw diagrams of plant transport tissues

An organ usually contains many different types of cell. These are arranged in a particular pattern characteristic of the organ, with cells of a similar type found together, forming distinctive tissues.

Drawing plan diagrams

A plan diagram shows the outline of the various tissues in an organ such as a leaf (Figure 7.1) or an eye. It does *not* show individual cells.

You can use an eyepiece graticule to help you to show the dimensions of the different tissues in the correct proportions. For example, for the leaf, you could count the number of eyepiece graticule divisions across the palisade layer, and the number across the spongy layer, so that you can draw these the correct sizes in relation to one another. You do not need to convert the eyepiece graticule divisions to real units (e.g. μm) so there is no need to use a stage micrometer.

Drawing cellular detail

Rather than a plan diagram, you may be asked to draw details of the cells you can see using a microscope. Use high power, and draw just a few representative cells, showing details of organelles if these are visible.

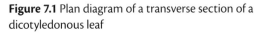

Figure 7.1 Plan diagram of a transverse section of a dicotyledonous leaf

 **Xylem tissue** Revised

Xylem tissue contains dead, empty cells with no end walls. These are called **xylem vessel elements**. They are arranged in long lines to form **xylem vessels** (Figure 7.2). These are long, hollow tubes through which water moves by mass flow from the roots to all other parts of the plant.

Transverse section

Cell wall containing cellulose and lignin — lignin makes the wall impermeable to water and provides strength, so the vessel element does not collapse when there is negative pressure inside it

Narrow lumen increases area of water in contact with wall; water molecules adhere to the walls and this helps to prevent breakage of the water column

Pit in cell wall allows movement of water out of the vessel element to other vessel elements or to neighbouring tissues

Longitudinal section

Dead cells have no contents, allowing easy movement of a continuous column of water by mass flow

Loss of end walls allows continuous movement of a column of water by mass flow

Figure 7.2 The structure of xylem vessels

Phloem tissue — Revised

Phloem tissue contains cells called **sieve tube elements**. Unlike xylem vessel elements, these are living cells and contain cytoplasm and a few organelles but no nucleus. Their walls are made of cellulose. A **companion cell** is associated with each sieve tube element (Figure 7.3).

Longitudinal section

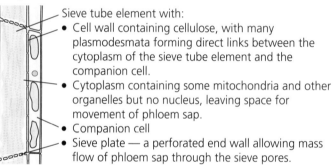

Sieve tube element with:
- Cell wall containing cellulose, with many plasmodesmata forming direct links between the cytoplasm of the sieve tube element and the companion cell.
- Cytoplasm containing some mitochondria and other organelles but no nucleus, leaving space for movement of phloem sap.
- Companion cell
- Sieve plate — a perforated end wall allowing mass flow of phloem sap through the sieve pores.

Transverse section

Companion cell with:
- Cytoplasm containing numerous organelles, including a nucleus and many mitochondria

Figure 7.3 Phloem tissue

Transport mechanisms

Transport in xylem — Revised

Figure 7.4 shows the pathway taken by water through a plant. The driving force that causes this movement is the loss of water vapour from the leaves. This is called **transpiration**.

Transpiration

Transpiration is the loss of water vapour from a plant. Most transpiration happens in the leaves. A leaf contains many cells in contact with air spaces in the mesophyll layers. Liquid water in the cell walls changes to water vapour, which diffuses into the air spaces. The water vapour then diffuses out of the leaf through the stomata, down a water potential gradient, into the air surrounding the leaf (Figure 7.5).

Each stoma is surrounded by a pair of guard cells. These can change shape to open or close the stoma. In order to photosynthesise, the stomata must be open so that carbon dioxide can diffuse into the leaf. Plants cannot therefore avoid losing water vapour by transpiration.

> **Transpiration** is the loss of water vapour from a plant to its environment, by diffusion down a water potential gradient, usually through the stomata in the leaf epidermis.

Now test yourself

1 Use the diagram at the bottom left in Figure 7.4 to draw a plan diagram of a transverse section of a root.

Answer on p.203

Tested

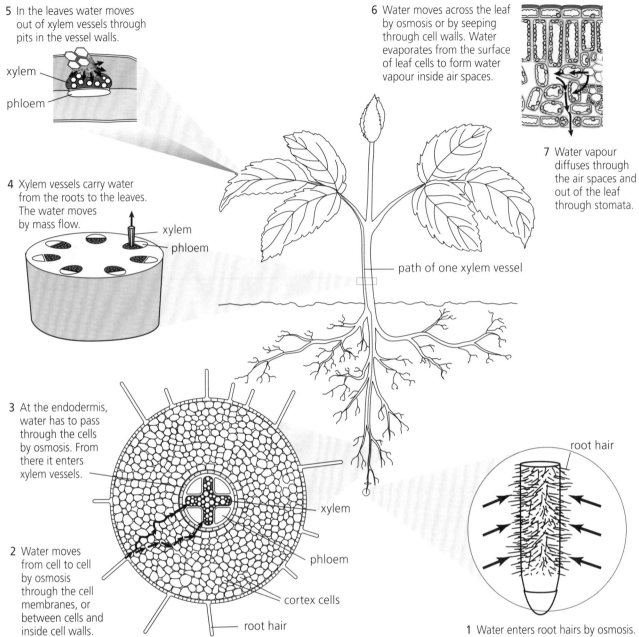

5 In the leaves water moves out of xylem vessels through pits in the vessel walls.

xylem

phloem

6 Water moves across the leaf by osmosis or by seeping through cell walls. Water evaporates from the surface of leaf cells to form water vapour inside air spaces.

7 Water vapour diffuses through the air spaces and out of the leaf through stomata.

4 Xylem vessels carry water from the roots to the leaves. The water moves by mass flow.

xylem

phloem

path of one xylem vessel

root hair

3 At the endodermis, water has to pass through the cells by osmosis. From there it enters xylem vessels.

xylem

phloem

cortex cells

2 Water moves from cell to cell by osmosis through the cell membranes, or between cells and inside cell walls.

root hair

1 Water enters root hairs by osmosis.

Figure 7.4 The pathway of water through a plant

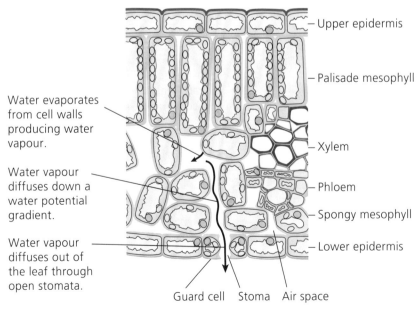

Upper epidermis

Palisade mesophyll

Water evaporates from cell walls producing water vapour.

Xylem

Water vapour diffuses down a water potential gradient.

Phloem

Spongy mesophyll

Water vapour diffuses out of the leaf through open stomata.

Lower epidermis

Guard cell   Stoma   Air space

Figure 7.5 Transpiration

Transpiration is affected by several factors:

- High temperature increases the rate of transpiration. This is because at higher temperatures water molecules have more kinetic energy. Evaporation from the cell walls inside the leaf therefore happens more rapidly, and diffusion also happens more rapidly.
- High humidity decreases the rate of transpiration. This is because the water potential gradient between the air spaces inside the leaf and the air outside is less steep, so diffusion of water vapour out of the leaf happens more slowly.
- High wind speed increases the rate of transpiration. This is because the moving air carries away water vapour from the surface of the leaf, helping to maintain a water potential gradient between the air spaces inside the leaf and the air outside.
- High light intensity may increase the rate of transpiration. This is because the plant may be photosynthesising rapidly, requiring a rapid supply of carbon dioxide. This means that more stomata are likely to be open, through which water vapour can diffuse out of the leaf.

How to... Investigate the factors that affect transpiration rate

It is difficult to measure the rate at which water vapour is lost from leaves. It is much easier to measure the rate at which a plant, or part of a plant, takes up water. Most of the water taken up is lost through transpiration, so we can generally assume that an increase in the rate of take-up of water indicates an increase in the rate of transpiration.

The apparatus used to measure the rate of take-up of water of a plant shoot is called a **potometer**. This can simply be a long glass tube. More complex potometers may have reservoirs, which make it easier to refill the tube with water, or a scale marked on them.

- Fix a short length of rubber tubing over one end of the long glass tube. Completely submerge the tube in water. Move it around to get rid of all air inside it and fill it with water. Make absolutely sure there are no air bubbles.
- Take a leafy shoot from a plant and submerge it in the water alongside the glass tube. Using a sharp blade, make a slanting cut across the stem.
- Push the cut end of the stem into the rubber tubing. Make sure the fit is tight and that there are no air bubbles. If

necessary, use a small piece of wire to fasten the tube tightly around the stem.

- Take the whole apparatus out of the water and support it upright. Wait at least 10 minutes for it to dry out. If the glass tube is not marked with a scale, place a ruler or graph paper behind it.
- Start a stop clock and read the position of the air/water meniscus (which will be near the base of the tube). Record its position every 2 minutes (or whatever time interval seems sensible). Stop when you have 10 readings, or when the meniscus is one third of the way up the tube.
- Change the environmental conditions and continue to take readings. For example, you could use a fan to increase 'wind speed', or move the apparatus into an area where the temperature is higher or lower.
- Plot distance moved by meniscus against time for each set of readings, on the same axes. Draw best fit lines. Calculate the mean distance moved per minute, or calculate the slope of each line. This can be considered to be the rate of transpiration.

How water moves from soil to air

Water moves from the soil to the air through a plant down a water potential gradient (Figure 7.6). The water potential in the soil is generally higher than in the air. The water potential in the leaves is kept lower than the water potential in the soil because of the loss of water vapour by transpiration. Transpiration maintains the water potential gradient.

- Water enters root hair cells by osmosis, moving down a water potential gradient from the water in the spaces between soil particles, through the cell surface membrane and into the cytoplasm and vacuole of the root hair cell.
- The water then moves from the root hair cell to a neighbouring cell by osmosis, down a water potential gradient. This is called the **symplastic** pathway.
- Water also seeps into the cell wall of the root hair cell. This does not involve osmosis, as no partially permeable membrane is crossed. The water then seeps into and along the cell walls of neighbouring cells. This is called the **apoplastic** pathway. In most plant roots, the apoplastic pathway carries more water than the symplastic pathway.

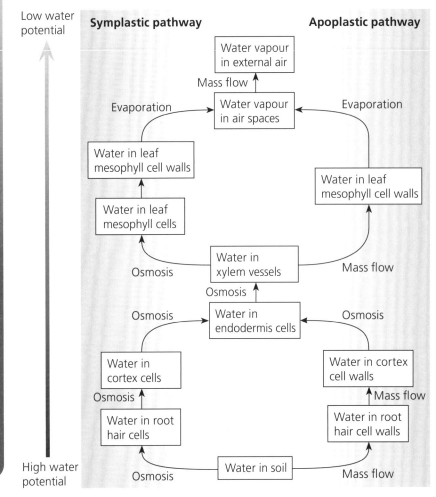

Low water
potential

High water
potential

Figure 7.6 Summary of water movement through a plant from soil to air

- When the water nears the centre of the root, it encounters a cylinder of cells called the **endodermis**. Each cell has a ring of impermeable **suberin** around it, forming the **Casparian strip**. This prevents water continuing to seep through cell walls. It therefore travels through these cells by the symplastic pathway.
- The water moves into the xylem vessels from the endodermis.
- Water moves up the xylem vessels by mass flow — that is, in a similar way to water flowing in a river. The water molecules are held together by hydrogen bonds between them, keeping the water column unbroken. This is called **cohesion**, and it helps to create a tension in the water column as it is drawn upwards. The water molecules are also attracted to the cellulose and lignin in the walls of the xylem vessels, by a force called **adhesion**. There is a relatively low hydrostatic pressure at the top of the column, produced by the loss of water by transpiration. This lowering of hydrostatic pressure causes a pressure gradient from the base to the top of the xylem vessel.
- In a leaf, water moves out of xylem vessels through pits, and then across the leaf by the apoplastic and symplastic pathways.
- Water evaporates from the wet cell walls into the leaf spaces, and then diffuses out through the stomata.

Xerophytes

A xerophyte is a plant that is adapted to live in an environment where water is in short supply. The adaptations may include:

- leaves with a small surface area-to-volume ratio. This reduces the amount of surface area from which water vapour can diffuse.
- leaves with a thick, waxy cuticle. This reduces the quantity of water that can diffuse through the surface of the leaf into the air.

- methods of trapping moist air near the stomata, for example rolling the leaf with the stomata inside, having stomata in pits in the leaf surface, having hairs around the stomata. This produces a layer of high water potential around the stomata, reducing the water potential gradient and therefore reducing the rate of diffusion of water vapour from inside the leaf to outside.

Transport in phloem
Revised

The movement of substances in phloem tissue is called **translocation**. The main substances that are moved are **sucrose** and **amino acids**, which are in solution in water. These substances have been made by the plant and are called **assimilates**.

How assimilates move through phloem tissue
A part of a plant where assimilates such as sucrose enter the phloem is called a **source**. A part where assimilates leave the phloem is called a **sink**. For example, a leaf may be a source and a root may be a sink.

Translocation of sucrose and other assimilates is an energy-requiring process.

- Respiration in companion cells at a source provides ATP that is used to fuel the transport of sucrose into the companion cell. This is done by pumping protons (hydrogen ions) out of the companion cell. As the protons diffuse back into the companion cell, they pass through a co-transporter protein that allows both protons and sucrose molecules to enter the cell. The protons carry sucrose molecules with them. This increases the concentration of sucrose in the companion cell, so that it moves by diffusion down a concentration gradient, through the plasmodesmata, into the phloem sieve element.
- The increased concentration of sucrose in the companion cell and phloem sieve element produces a water potential gradient from the surrounding cells into the companion cell and phloem sieve element. Water moves down this gradient.
- At a sink, sucrose diffuses out of the phloem sieve element and down a concentration gradient into a cell that is using sucrose. This produces a water potential gradient, so water also diffuses out of the phloem sieve element.
- The addition of water at the source and the loss of water at the sink produces a higher hydrostatic pressure inside the phloem sieve element at the source than at the sink. Phloem sap therefore moves by **mass flow** down this pressure gradient, through the phloem sieve elements and through the sieve pores, from source to sink.

> **Revision activity**
>
> - Construct a table comparing the structures and functions of xylem vessels and phloem sieve tubes.

Now test yourself
Tested

2 At which of these stages of transport of sucrose does the plant have to provide energy?
  a loading sucrose into the phloem sieve tube
  b mass flow of phloem sap from source to sink
3 Explain why the contents of xylem vessels always flow upwards from roots to leaves, while the contents of phloem sieve tubes can flow either upwards or downwards.
4 Plants contain many different organs, including flowers, leaves, roots and fruits. In a temperate country, different organs in a plant are sources and sinks at different times of year. For each of the following, state which of these organs will be sources and which will be sinks:
  a in summer, when there is plenty of sunlight and flower buds are beginning to open
  b in autumn, when there is still plenty of sunlight and the plant is building up stores of starch in its roots ready for the winter; flowers have been fertilised and fruits are developing
  c in winter, when there is little sunlight, and the plant relies on starch stores in its roots
  d in spring, when the leaves are just starting to grow

Answers on p.203

The circulatory system

Mammals have a blood system made up of blood vessels and the heart. The heart produces high pressure, which causes blood to move through the vessels by mass flow. The blood system is called a **closed system** because the blood travels inside vessels. It is called a **double circulatory system** because the blood flows from the heart to the lungs, then back to the heart, then around the rest of the body and then back to the heart again. The vessels taking blood to the lungs and back make up the **pulmonary system**. The vessels taking blood to the rest of the body and back make up the **systemic system**.

> **Expert tip**
>
> Mass flow means that a whole body of liquid flows together, like a river. It is much faster than diffusion, where individual molecules or ions move randomly.

Blood vessels   Revised ☐

Arteries carry blood away from the heart. The blood that flows through them is pulsing and at a high pressure. They therefore have thick, elastic walls, which can expand and recoil as the blood pulses through. The artery wall also contains variable amounts of smooth muscle. Arteries branch into smaller vessels called **arterioles**. These also contain smooth muscle in their walls, which can contract and make the lumen (space inside) smaller. This helps to control the flow of blood to different parts of the body.

> **Typical mistake**
>
> Do not state that the muscle in artery walls pumps blood through them – this is not correct.

Capillaries are tiny vessels with just enough space for red blood cells to squeeze through. Their walls are only one cell thick, and there are often gaps in the walls through which plasma (the liquid component of blood) can leak out. Capillaries deliver nutrients, hormones and other requirements to body cells, and take away their waste products. Their small size and thin walls minimise diffusion distance, enabling exchange to take place rapidly between the blood and the body cells.

Veins carry low-pressure blood back to the heart. Their walls do not need to be as tough or as elastic as those of arteries as the blood is not at high pressure and is not pulsing. The lumen is larger than in arteries (Figure 8.1), reducing friction, which would otherwise slow down blood movement. They contain valves, to ensure that the blood does not flow back the wrong way. Blood is kept moving through many veins, for example those in the legs, by the squeezing effect produced by contraction of the body muscles close to them, which are used when walking.

Figure 8.2 shows the main blood vessels in the human body.

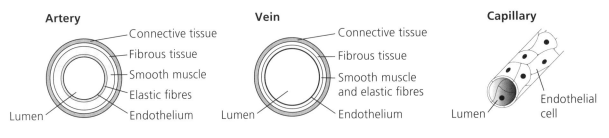

Figure 8.1 Structure of blood vessels

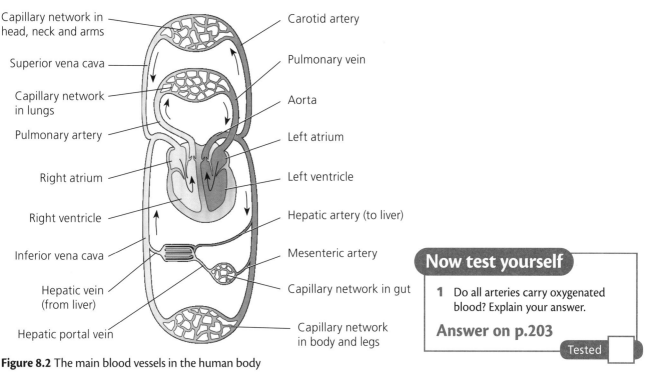

Capillary network in head, neck and arms

Superior vena cava

Capillary network in lungs

Pulmonary artery

Right atrium

Right ventricle

Inferior vena cava

Hepatic vein (from liver)

Hepatic portal vein

Carotid artery

Pulmonary vein

Aorta

Left atrium

Left ventricle

Hepatic artery (to liver)

Mesenteric artery

Capillary network in gut

Capillary network in body and legs

Figure 8.2 The main blood vessels in the human body

Now test yourself

1 Do all arteries carry oxygenated blood? Explain your answer.

Answer on p.203

Tested

Pressure changes in the circulatory system

Revised

The pressure of the blood changes as it moves through the circulatory system (Figure 8.3).

● In the arteries, blood is at high pressure because it has just been pumped out of the heart. The pressure oscillates (goes up and down) in time with the heart beat. The stretching and recoil of the artery walls helps to smooth the oscillations, so the pressure becomes gradually steadier the further the blood moves along the arteries. The mean pressure also gradually decreases, particularly as the blood flows through arterioles (small arteries).

● The total cross-sectional area of the capillaries is greater than that of the arteries that supply them, so blood pressure is less inside the capillaries than inside arteries.

● In the veins, blood is at a very low pressure, as it is now a long way from the pumping effect of the heart.

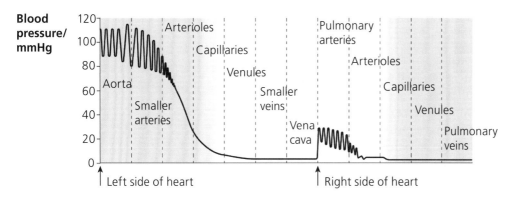

Blood pressure/ mmHg

Figure 8.3 Pressure changes in the circulatory system

Blood

Blood contains the following cell types (Figure 8.4).

- **Red blood cells**. These transport oxygen from lungs to respiring tissues. They are very small and have the shape of a biconcave disc. This increases their surface area-to-volume ratio, allowing rapid diffusion of oxygen into and out of them. They contain haemoglobin, which combines with oxygen to form oxy-haemoglobin in areas of high concentration (such as the lungs) and releases oxygen in areas of low concentration (such as respiring tissues). They do not contain a nucleus or mitochondria.
- **Phagocytes**. These are white blood cells that engulf and digest unwanted cells, such as damaged body cells or pathogens. They are larger than red blood cells. They include monocytes and neutrophils.
- **Lymphocytes**. These are white blood cells that respond to particular pathogens by secreting antibodies or by directly destroying them. Each lymphocyte is able to recognise one particular pathogen and respond to it by secreting one particular type of antibody or by attacking it. You can find out more about this on pp. 69–71.

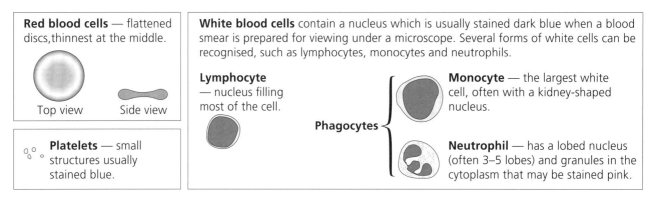

Red blood cells — flattened discs, thinnest at the middle. Top view Side view **Platelets** — small structures usually stained blue.	**White blood cells** contain a nucleus which is usually stained dark blue when a blood smear is prepared for viewing under a microscope. Several forms of white cells can be recognised, such as lymphocytes, monocytes and neutrophils. **Lymphocyte** — nucleus filling most of the cell. **Phagocytes** **Monocyte** — the largest white cell, often with a kidney-shaped nucleus. **Neutrophil** — has a lobed nucleus (often 3–5 lobes) and granules in the cytoplasm that may be stained pink.

Figure 8.4 Blood cells

Haemoglobin and oxygen transport

Haemoglobin (Hb) is a protein with quaternary structure. A haemoglobin molecule is made up of four polypeptide chains, each of which has a **haem** group at its centre. Each haem group contains an Fe^{2+} ion, which is able to combine reversibly with oxygen, forming **oxyhaemoglobin**. Each iron ion can combine with two oxygen atoms, so one Hb molecule can combine with eight oxygen atoms.

Oxygen concentration can be measured as partial pressure, in kilopascals (kPa). Haemoglobin combines with more oxygen at high partial pressures than it does at low partial pressures. At high partial pressures of oxygen, all the haemoglobin will be combined with oxygen, and we say that it is 100% saturated with oxygen. A graph showing the relationship between the partial pressure of oxygen and the percentage saturation of haemoglobin with oxygen is known as a dissociation curve (Figure 8.5). In the lungs, the partial pressure of oxygen may be around 12 kPa. You can see from the graph that the Hb will be about 98% saturated.

In a respiring muscle, the partial pressure of oxygen may be around 2 kPa. The Hb will be about 23% saturated.

Therefore, when Hb from the lungs arrives at a respiring muscle it gives up more than 70% of the oxygen it is carrying.

Expert tip

When explaining how the structure of haemoglobin is related to its functions, remember that its ability to unload oxygen is just as important as its ability to combine with it.

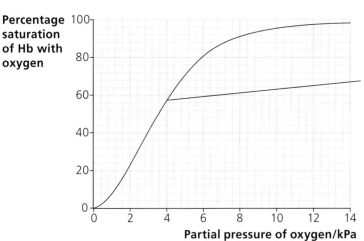

Figure 8.5 The oxygen dissociation curve for haemoglobin

The steepest part of the curve indicates where relatively small changes in partial pressure of oxygen cause the greatest loading or unloading of Hb with oxygen.

The Bohr effect

The presence of carbon dioxide increases acidity, that is, the concentration of H^+ ions. When this happens, the haemoglobin combines with H^+ ions and releases oxygen.

Red blood cells contain an enzyme called **carbonic anhydrase**, which catalyses the reaction of carbon dioxide and water to form carbonic acid:

$$CO_2 + H_2O \xrightleftharpoons[]{\text{carbonic anhydrase}} H_2CO_3$$

The carbonic acid then dissociates:

$$H_2CO_3 \rightleftharpoons H^+ + HCO_3^-$$

The hydrogen ions combine with haemoglobin to form **haemoglobinic acid**. This causes the haemoglobin to release oxygen.

Therefore, in areas of high carbon dioxide concentration, haemoglobin is less saturated with oxygen than it would be if there was no carbon dioxide present. This is called the **Bohr effect** (Figure 8.6). It is useful in enabling haemoglobin to unload more of its oxygen in tissues where respiration (which produces carbon dioxide) is taking place.

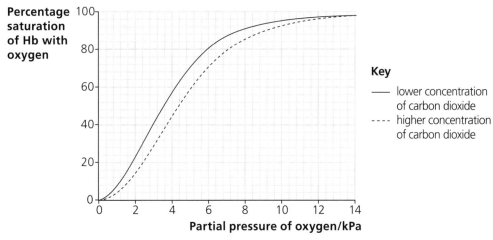

Figure 8.6 The Bohr effect

Transport of carbon dioxide

Carbon dioxide is transported in the blood in three different ways:

- 85% as hydrogencarbonate ions, HCO_3^-, as described above
- 10% combined with haemoglobin, to form **carbaminohaemoglobin**
- 5% in solution in blood plasma

Adaptation to high altitude

At high altitudes, the air is less dense and the partial pressure of oxygen is lower than at sea level. Haemoglobin is therefore less saturated with oxygen in the lungs and delivers less oxygen to body tissues.

After some time at high altitude, the number of red blood cells in the blood increases. This means that there are more haemoglobin molecules in a given volume of blood. Therefore, even though each Hb molecule carries less oxygen on average than at sea level, the fact that there are more of them helps to supply the same amount of oxygen to respiring tissues.

Athletes can make use of this by training at high altitude before an important competition. When they return to low altitude, their extra red blood cells can supply oxygen to their muscles at a greater rate than in an athlete who has not been to high altitude, giving them a competitive advantage.

Tissue fluid and lymph
Revised

Capillaries have tiny gaps between the cells in their walls. Near the arteriole end of a capillary, there is relatively high pressure inside the capillary, and plasma leaks out through these gaps to fill the spaces between the body cells. This leaked plasma is called **tissue fluid**.

Tissue fluid is therefore very similar to blood plasma. However, very large molecules such as albumin (a protein carried in solution in blood plasma) and other plasma proteins cannot get through the pores and so remain in the blood plasma.

The tissue fluid bathes the body cells. Substances such as oxygen, glucose or urea can move between the blood plasma and the cells by diffusing through the tissue fluid.

Some tissue fluid moves back into the capillaries, becoming part of the blood plasma once more. This happens especially at the venule end of the capillary where blood pressure is lower, producing a pressure gradient down which the tissue fluid can flow. However, some of the tissue fluid collects into blind-ending vessels called lymphatic vessels. It is then called **lymph**.

Lymphatic vessels have valves that allow fluid to flow into them and along them but not back out again. They carry the lymph towards the subclavian veins (near the collarbone) where it is returned to the blood. The lymph passes through **lymphatic glands** where white blood cells accumulate. Lymph therefore tends to carry higher densities of white blood cells than are found in blood plasma or tissue fluid.

Revision activity
- Construct a table to compare blood, tissue fluid and lymph.

The heart

The heart of a mammal (Figure 8.7) has four chambers. The two **atria** receive blood, and the two **ventricles** push blood out of the heart (Figure 8.8). The atria and ventricle on the left side of the heart contain oxygenated blood, while those on the right side contain deoxygenated blood. The walls of the heart are made of **cardiac muscle**.

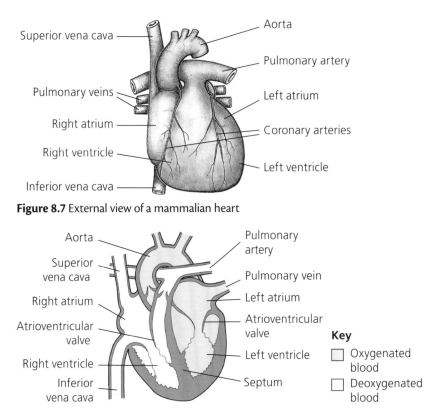

Figure 8.7 External view of a mammalian heart

Key
☐ Oxygenated blood
☐ Deoxygenated blood

Note: the walls of the ventricles are thicker than those of the atria because the cardiac muscle has to produce more pressure. The wall of the left ventricle is thicker than the wall of the right ventricle because the right ventricle only has to push blood through the capillaries of the lungs, where the resistance to flow is much less than through all the capillaries of the body organs supplied with blood from the left ventricle.

Figure 8.8 Vertical section through a mammalian heart

The cardiac cycle

Revised ☐

When muscle contracts, it gets shorter. Contraction of the cardiac muscle in the walls of the heart therefore causes the walls to squeeze inwards on the blood inside the heart. Both sides of the heart contract and relax together. The complete sequence of one heart beat is called the **cardiac cycle** (Figures 8.9 and 8.10).

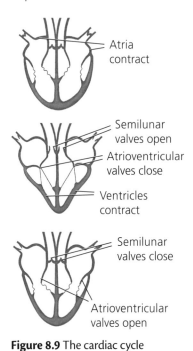

During **atrial systole**, the muscle in the walls of the atria contracts, pushing more blood into the ventricles through the open atrioventricular valves.

During **ventricular systole**, the muscle in the walls of the ventricles contracts. This causes the pressure of the blood inside the ventricles to become greater than in the atria, forcing the atrioventricular valves shut. The blood is forced out through the aorta and pulmonary artery.

During **diastole**, the heart muscles relax. The pressure inside the ventricles becomes less than that inside the aorta and pulmonary artery, so the blood inside these vessels pushes the semilunar valves shut. Blood flows into the atria from the veins, so the cycle is ready to begin again.

Expert tip

Remember that the valves are opened and closed by differences in blood pressure. They cannot move themselves.

Figure 8.9 The cardiac cycle

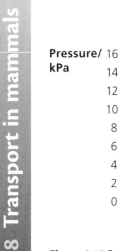

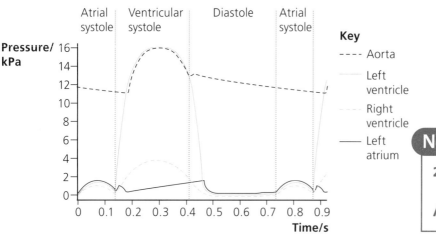

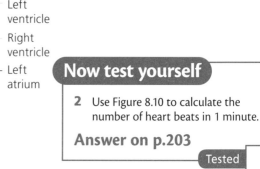

Now test yourself

2 Use Figure 8.10 to calculate the number of heart beats in 1 minute.

Answer on p.203

Tested

Figure 8.10 Pressure changes during the cardiac cycle

Initiation and control of the cardiac cycle

Cardiac muscle is **myogenic** — that is, it contracts and relaxes automatically, without the need for stimulation by nerves. The rhythmic contraction of the cardiac muscle in different parts of the heart is coordinated by electrical impulses passing through the cardiac muscle tissue.

● In the wall of the right atrium, there is a patch of muscle tissue called the **sinoatrial node** (**SAN**). This has an intrinsic rate of contraction a little higher than that of the rest of the heart muscle.

● As the cells in the SAN contract, they generate action potentials (electrical impulses — see p. 130) which sweep along the muscle in the wall of the right and left atria. This causes the muscle to contract. This is atrial systole.

● When the action potentials reach the **atrioventricular node** (**AVN**) in the septum, they are delayed briefly. They then sweep down the septum between the ventricles, along fibres of **Purkyne tissue**, and then up through the ventricle walls. This causes the ventricles to contract slightly after the atria. The left and right ventricles contract together, from the bottom up. This is ventricular systole.

● There is then a short delay before the next wave of action potentials is generated in the SAN. During this time, the heart muscles relax. This is diastole.

9 Gas exchange and smoking

All organisms take in gases from their environment and release gases to the environment. Animals take in oxygen for aerobic respiration and release carbon dioxide. Plants also respire, but during daylight hours they photosynthesise at a greater rate than they respire, and so take in carbon dioxide and release oxygen.

The body surface across which these gases diffuse into and out of the body is called the gas exchange surface. In mammals, including humans, the gas exchange surface is the surface of the alveoli in the lungs.

> **Gas exchange** is the movement of gases into and out of an organism's body, across a gas exchange surface.

The human gas exchange system

Structure of the human gas exchange system
Revised

Figure 9.1 shows the structure of the human gas exchange system.

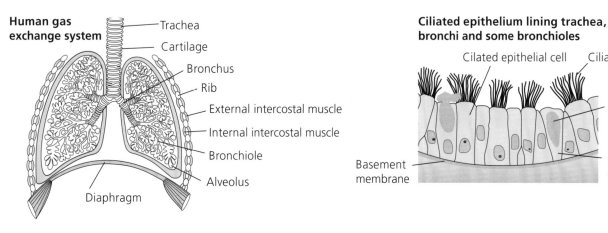

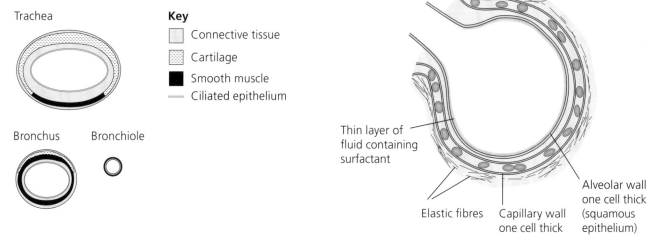

Figure 9.1 The structure of the human gas exchange system

Cartilage in the walls of the trachea and bronchi provides support and prevents the tubes collapsing when the air pressure inside them is low.

Ciliated epithelium is found lining the trachea, bronchi and some bronchioles. It is a single layer of cells whose outer surfaces are covered with many thin extensions (cilia), which are able to move. They sweep mucus upwards towards the mouth, helping to prevent dust particles and bacteria reaching the lungs.

Goblet cells are also found in the ciliated epithelium. They secrete mucus, which traps dust particles and bacteria.

Smooth muscle cells are found in the walls of the trachea, bronchi and bronchioles. This type of muscle can contract slowly but for long periods without tiring. When it contracts, it reduces the diameter of the tubes. During exercise it relaxes, widening the tubes so more air can reach the lungs.

Elastic fibres are found in the walls of all tubes and between the alveoli. When breathing in, these fibres stretch to allow the alveoli and airways to expand. When breathing out, they recoil, helping to reduce the volume of alveoli and expel air out of the lungs.

Gas exchange at the alveolar surface
Revised

The air inside an alveolus contains a higher concentration of oxygen, and a lower concentration of carbon dioxide, than the blood in the capillaries. This blood has been brought to the lungs in the pulmonary artery, which carries deoxygenated blood from the heart. Oxygen therefore diffuses from the alveolus into the blood capillary, through the thin walls of the alveolus and the capillary. Carbon dioxide diffuses from the capillary into the alveolus.

The diffusion gradients for these gases are maintained by:
- breathing movements, which draw air from outside the body into the lungs, and then push it out again; this maintains a relatively high concentration of oxygen and low concentration of carbon dioxide in the alveoli
- blood flow past the alveolus, which brings deoxygenated blood and carries away oxygenated blood

Cigarette smoking and health

The smoke from cigarettes contains several substances that affect the gas exchange system and the cardiovascular system. These include:
- **tar**, a mixture of substances including various chemicals that act as **carcinogens**.
- **nicotine**, an addictive substance that affects the nervous system by binding to receptors on neurones (nerve cells) in the brain and other parts of the body. It increases the release of a neurotransmitter called dopamine in the brain, which gives feelings of pleasure. It increases the release of adrenaline into the blood, which in turn increases breathing rate and heart rate. There is also some evidence that nicotine increases the likelihood of blood clots forming.
- **carbon monoxide**, which combines irreversibly with haemoglobin, forming carboxyhaemoglobin. This reduces the amount of haemoglobin available to combine with oxygen, and so reduces the amount of oxygen that is transported to body tissues.

Typical mistake

Do not confuse carboxyhaemoglobin with carbaminohaemoglobin (see p. 5).

Effects of smoking on the gas exchange system

Revised

Chronic obstructive pulmonary disease (COPD)

This is a condition in which a person has chronic bronchitis and emphysema. It can be extremely disabling, as it is difficult to get sufficient oxygen into the blood to support activity.

Chronic bronchitis

Various components of cigarette smoke, including tar, cause goblet cells to increase mucus production and cilia to beat less strongly. This causes mucus to build up, which may partially block alveoli. This makes gas exchange more difficult, as the diffusion distance between the air in the alveoli and the blood in the capillaries is greater. The mucus may become infected with bacteria, causing **bronchitis**. Smokers often have chronic (long-lasting) bronchitis.

The mucus stimulates persistent coughing, which can damage the tissues in the walls of the airways, making them stiffer and the airways narrower.

Emphysema

Smoking causes inflammation in the lungs. This involves the presence of increased numbers of white blood cells, some of which secrete chemicals that damage elastic fibres. This makes the alveoli less elastic. They may burst, resulting in larger air spaces. This reduces the surface area available for gas exchange. This is called **emphysema**. A person with emphysema has shortness of breath, meaning they struggle to breathe as deeply as they need to, especially when exercising.

Lung cancer

Various components of tar can cause changes in the DNA in body cells, including the genes that control cell division, which can cause cancer. These substances are therefore **carcinogens**. Cancers caused by cigarette smoke are most likely to form in the lungs but may form anywhere in the gas exchange system, and also in other parts of the body. Smoking increases the risk of developing all types of cancer. Symptoms of lung cancer include shortness of breath, a chronic cough — which may bring up blood — chest pain, fatigue and weight loss.

Effects of smoking on the cardiovascular system

Revised

The nicotine and carbon monoxide in tobacco smoke increase the risk of developing **atherosclerosis**. Atherosclerosis is a thickening and loss of elasticity in the walls of arteries. It is caused by build-up of plaques in the blood vessel wall. The plaques contain cholesterol and fibres. They produce a rough surface lining the artery, which stimulates the formation of blood clots.

A blood clot may break away from the artery wall and get stuck in a narrow vessel elsewhere in the blood system, for example in the lungs or in the brain. This prevents blood passing through so cells are not supplied with oxygen and die. If this happens in the brain it is called a **stroke**.

The loss of elasticity in an artery or arteriole also makes it more likely that the vessel will burst when high-pressure blood pulses through. This is another cause of stroke.

If atherosclerosis happens in the coronary arteries that supply the heart muscle with oxygenated blood, the person has **coronary heart disease** (CHD). Parts of the muscle may be unable to function properly as they do not have enough oxygen for aerobic respiration. The muscle may die. Eventually, this part of the heart might stop beating, causing a heart attack.

Revision activity

● Construct a spider diagram to show the different components of cigarette smoke, and the effects that they have on health.

10 Infectious disease

Disease can be defined as a condition in which the body does not function normally, and which produces unpleasant symptoms such as pain, distress or feeling weak. The term disease is generally used for conditions that last for at least several days.

> A **disease** is a disorder of the body that leads to ill health.

An **infectious disease** is one that can be passed between one person and another. Infectious diseases are caused by **pathogens**. These are usually microorganisms such as viruses, bacteria, fungi or protoctists.

A **non-infectious disease** cannot be passed between people and is not caused by pathogens. Examples include sickle cell anaemia and lung cancer.

Important infectious diseases

Cholera
Revised ☐

Cause
Cholera is caused by a bacterium, *Vibrio cholerae*.

> The **cause** of an infectious disease is the pathogen that produces the symptoms of the disease.

Transmission
V. cholerae can enter the body in contaminated food or water. The bacteria breed in the small intestine, where they secrete a toxin that reduces the ability of the epithelium of the intestine to absorb salts and water into the blood. These are lost in the faeces, causing diarrhoea. If not treated, the loss of fluid can be fatal. Cholera is most likely to occur where people use water or food that has been in contact with untreated sewage, as the bacteria are present in the faeces of an infected person.

> **Transmission** is the movement of a pathogen from an infected person to an uninfected person.

Prevention and control
Transmission is most likely to occur in crowded and impoverished conditions, such as refugee camps. Cholera is best controlled by treating sewage effectively, providing a clean water supply and maintaining good hygiene in food preparation. There is no fully effective vaccine for cholera.

Malaria
Revised ☐

Cause
Malaria is caused by a protoctist, *Plasmodium*. There are several species, which cause different types of malaria. In a person, the *Plasmodium* infects red blood cells and breeds inside them. Toxins are released when the *Plasmodium* burst out of the cells, causing fever.

Transmission

Plasmodium is transmitted in the saliva of female *Anopheles* mosquitoes, which inject saliva to prevent blood clotting when they feed on blood from a person. When a mosquito bites an infected person, *Plasmodium* is taken up into the mosquito's body and eventually reaches its salivary glands. The mosquito is said to be a **vector** for malaria.

Prevention and control

Reducing the population of mosquitoes, for example by removing sources of water in which they can breed, or by releasing large numbers of sterile males, can reduce the transmission of malaria.

Preventing mosquitoes from biting people, for example by sleeping under a mosquito net, or by wearing long-sleeved clothing and insect repellant, can reduce the chances of a mosquito picking up *Plasmodium* from an infected person, or passing it to an uninfected person.

Prophylactic drugs (that is, drugs that prevent pathogens infecting and breeding in a person) can be taken. However, in many parts of the world *Plasmodium* has evolved resistance to some of these drugs.

A **vector** is an organism that transmits a pathogen from one host to another.

Typical mistake

Mosquitoes are not the cause of malaria. Malaria is caused by *Plasmodium*. Mosquitoes are the vector for this disease.

Now test yourself

1 Explain the difference between the cause of an infectious disease, and a vector for an infectious disease.

Answer on p.203

Tested

Tuberculosis (TB)

Revised

Cause

TB is caused by the bacterium *Mycobacterium tuberculosis* or (more rarely) *Mycobacterium bovis*.

Transmission

M. tuberculosis can enter the lungs in airborne droplets of liquid that are breathed in. This is more likely to happen in places where many people are living in crowded conditions.

Prevention and control

TB is most prevalent amongst people living in poor accommodation, or whose immune systems are not functioning well, perhaps because of malnutrition or infection with HIV (see below). Increasing standards of living and treating HIV infection can therefore help to reduce the incidence of TB.

Vaccination with the BCG vaccine confers immunity to TB in many people. New vaccines are being developed that it is hoped will be more effective.

Treatment of HIV with drug therapy reduces the risk that an HIV-positive person will get TB.

Treatment of TB with antibiotics can often completely cure the disease. However, this is not always the case because:

● there are now many strains of the *M. tuberculosis* bacterium that have evolved resistance to most of the antibiotics that are used
● the bacteria reproduce *inside* body cells, where it is difficult for drugs to reach them
● the drugs need to be taken over a long time period, which often requires a health worker checking that a person takes their drugs every day

HIV/AIDS

Revised

Cause

AIDS (acquired immunodeficiency syndrome) is caused by the human immunodeficiency virus, HIV. This is a retrovirus, which contains RNA. The virus enters T-lymphocytes (p. 71), where its RNA is used to make viral DNA, which is incorporated into the T-lymphocytes' chromosomes. A person who has been infected with HIV makes antibodies against the virus, and is said to be HIV-positive. Usually nothing more happens for several years after infection, but eventually multiple copies of the virus are made inside the T-lymphocytes, which are destroyed as the viruses break out and infect more cells. Eventually there are so few functioning T-lymphocytes that the person is no longer able to resist infection by other pathogens and develops one or more opportunistic diseases such as TB. The person has AIDS.

Transmission

HIV can be passed from one person to another through:

- blood from one person entering that of another, for example by sharing hypodermic needles, or through blood transfusions
- exchange of fluids from the penis, vagina or anus
- across the placenta from mother to fetus, or in breast milk

Prevention and control

All blood to be used in transfusions should be screened to ensure it does not contain HIV.

All hypodermic needles should be sterile and used only once, and disposed of carefully.

A person should avoid sexual activity with anyone whose HIV status they do not know. If everyone had only one partner, HIV could not be transmitted. Condoms, if properly used, can prevent the virus passing from one person to another during intercourse. If a person is diagnosed with HIV, all their sexual contacts should be traced and informed that they may have the virus.

The chance of HIV passing from an HIV-positive mother to her fetus is greatly reduced if the mother is treated with appropriate drugs. These drugs can also greatly increase the length of time between a person becoming infected with HIV and developing symptoms of AIDS, and can significantly prolong life.

HIV infection rates are especially high in sub-Saharan Africa. Many of these people are not able to receive treatment with effective drugs, generally for economic reasons.

> **Now test yourself**
>
> 2 Explain the difference between being HIV-positive and having AIDS.
>
> **Answer on p.203**
>
> Tested

Smallpox

Revised

Cause

Smallpox is caused by the variola virus.

Transmission

Transmission occurs by the inhalation of droplets of moisture containing the virus.

Prevention and control

Smallpox was a serious disease that was fatal in 20–60% of adults who caught it, and in an even higher percentage of infected children. It was eradicated by 1979, through a vaccination campaign coordinated by the World Health Organization.

Measles

Revised

Cause

Measles is caused by a morbillivirus.

Transmission

Transmission occurs through the inhalation of droplets of moisture containing the virus. It is highly infectious, so a high proportion of people who come into close contact with an infected person will also get the disease.

Prevention and control

Measles is a serious disease, which can cause death, especially in adults and in people who are not in good health, for example because they do not have access to a good diet. Vaccination is the best defence against measles. The vaccine is highly effective, especially if two doses are given. The people most likely to suffer from measles are therefore those who are malnourished and who live in areas where no vaccination programme is in place.

Global patterns of disease

Revised

Malaria is found in parts of the world where the *Anopheles* mosquito species that can act as vectors are found. This is mostly in tropical and subtropical regions where humidity is high (Figure 10.1).

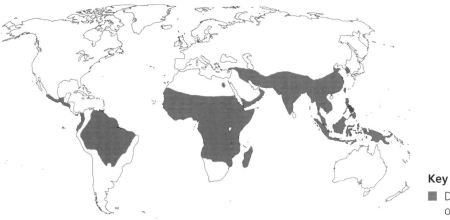

Key

■ Distribution of malaria

Figure 10.1 Global distribution of malaria

TB is found in all countries of the world, including developed countries such as the USA and the UK (Figure 10.2). However, it is most common in areas where living conditions are poor and crowded, or where large numbers of people have HIV/AIDS (Figure 10.3).

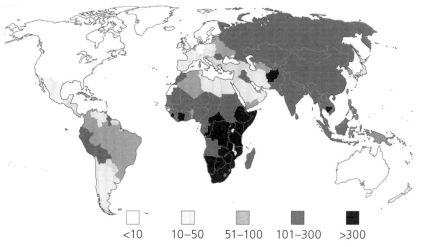

Key

Cases per 100 000 per year

| <10 | 10–50 | 51–100 | 101–300 | >300 |

Figure 10.2 Global distribution of TB

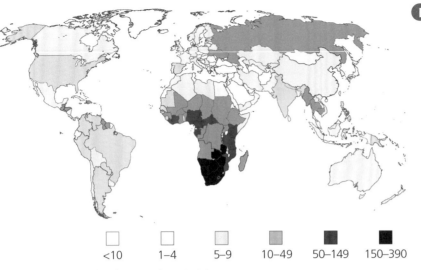

Revision activity

- Construct a table that shows the cause, method of transmission, and prevention and control methods for each of the infectious diseases described on pp. 64–67.

Key
Cases per 1000 people per year

| <10 | 1–4 | 5–9 | 10–49 | 50–149 | 150–390 |

Figure 10.3 Global distribution of HIV/AIDS

Antibiotics

How antibiotics work
Revised

An antibiotic is a substance that, when taken orally or by injection, kills bacteria but does not harm human cells. Antibiotics are not effective against viruses. Many antibiotics are originally derived from fungi, but they can also be obtained from other organisms (for example, amphibian skin or plants) or synthesised in the laboratory.

Antibiotics act on structures or metabolic pathways that are found in bacteria but not in eukaryotic cells. The correct antibiotic must be chosen for a particular disease. For example:

- **penicillin** prevents the synthesis of the links between peptidoglycan molecules in bacterial cell walls; when the bacteria take up water by osmosis, the cell wall is not strong enough to prevent them bursting
- **rifampicin (rifampin)** inhibits an enzyme required for RNA synthesis in bacteria
- **tetracycline** binds to bacterial ribosomes and inhibits protein synthesis

Resistance to antibiotics
Revised

Exposure to antibiotics exerts strong selection pressure on bacterial populations. Any bacterium that is resistant to the antibiotic — for example, because it has an allele of a gene that causes the synthesis of an enzyme that can break down the antibiotic — has a selective advantage and is more likely to survive and reproduce successfully. The offspring will inherit the alleles that confer resistance. A whole population of resistant bacteria can therefore be produced.

For this reason, it is important that antibiotics are only used when necessary. A person prescribed antibiotics should complete the course, as this increases the chances of eradicating all the disease-causing bacteria in the body. Using more than one antibiotic also reduces the chance of resistance developing, because it is unlikely that any one bacterium will possess two different resistance alleles.

11 Immunity

The immune system

The human immune system is made up of the organs and tissues involved in destroying pathogens inside the body. There are two main groups of cells involved:

- phagocytes, which ingest and digest pathogens or infected cells
- lymphocytes, which recognise specific pathogens through interaction with receptors in their cell surface membranes, and respond in one of several ways, for example by secreting antibodies

The numbers of both of these types of white blood cells increase when a person has an infectious disease, because mitosis of lymphocytes (see below) occurs as a result of contact with a pathogen. A type of cancer, called leukaemia, results from uncontrolled division of the stem cells that produce white blood cells, and this also causes an increase in their numbers. This can mean that the blood no longer has enough red blood cells, so oxygen transport is impaired.

Phagocytes
Revised

Phagocytes are produced in the bone marrow by the mitotic division of precursor cells. This produces cells that develop into **monocytes** or **neutrophils** (Figure 11.1).

Monocytes are inactive cells that circulate in the blood. They eventually leave the blood, often as the result of encountering chemical signals indicating that bacteria or viruses are present. As monocytes mature, they develop more RER, Golgi bodies and lysosomes. When they leave the blood they become **macrophages**. They engulf bacteria by endocytosis (p. 38) and digest them inside phagosomes. Monocytes and macrophages can live for several months.

Similar precursor cells in bone marrow produce **neutrophils**. These also travel in blood. They leave the blood in large numbers at sites of infection and engulf and digest bacteria in a similar way to macrophages. A neutrophil lives for only a few days.

 Monocyte    Neutrophil

Figure 11.1 Mature phagocytes

Phagocytes are able to act against any invading organisms. Their response is non-specific.

The immune response
Revised

Lymphocytes, unlike phagocytes, act against specific pathogens. Each lymphocyte contains a set of genes that codes for the production of a particular type of receptor. We have many million different types, each producing just one type of receptor.

Both **B-lymphocytes** and **T-lymphocytes** are made in bone marrow. B-lymphocytes then spread through the body and settle in lymph nodes, although some continue to circulate in the blood. T-lymphocytes collect in the thymus gland, where they mature before spreading into the same areas as B-lymphocytes. The thymus gland disappears at around the time of puberty. Both types of lymphocyte have a large, rounded nucleus that takes up most of the cell. They can only be told apart by their different actions (see below).

During the maturation process, any lymphocytes that produce receptors that would bind with those on the body's own cells are destroyed. This means that the remaining lymphocytes will only act against **non-self** molecules that are not normally found in the body. Non-self molecules, such as those on the surfaces of invading bacteria, are called **antigens**.

Several different types of cell, including macrophages, place antigens of pathogens they have encountered in their cell surface membranes, where there is a good chance that a B-lymphocyte or T-lymphocyte may encounter them. These cells are called **antigen-presenting cells**.

> An **antigen** is any molecule, usually a protein or glycoprotein, that is different from the molecules in the cell-surface membranes of our own body cells and therefore elicits an immune response.

Action of B-lymphocytes

A B-lymphocyte places some of its specific receptor molecules in its cell surface membrane. If it encounters an antigen that binds with this receptor, the B-lymphocyte is activated. It divides repeatedly by mitosis to produce a clone of genetically identical **plasma cells**. Some of these synthesise and secrete large quantities of proteins called **immunoglobulins** or **antibodies**. The antibodies have the same binding sites as the specific receptors in the B-lymphocyte's membrane, so they can bind with the antigens. This may directly destroy or neutralise the antigens, or it may make it easier for phagocytes to destroy them.

Some of the clone of B-lymphocyte cells become **memory cells**. These remain in the blood for many years. They are able to divide rapidly to produce plasma cells if the same antigen invades the body again. More antibody is therefore secreted more rapidly than when the first invasion happened, and it is likely that the pathogens will be destroyed before they have a chance to reproduce. The person has become **immune** to this pathogen (Figure 11.2).

> An **antibody** is a protein molecule secreted by a plasma cell that binds with specific antigens.

Now test yourself

1 B-lymphocytes have large numbers of mitochondria, extensive rough endoplasmic reticulum and several Golgi bodies. Explain why this is so.

Answer on p.203

Tested

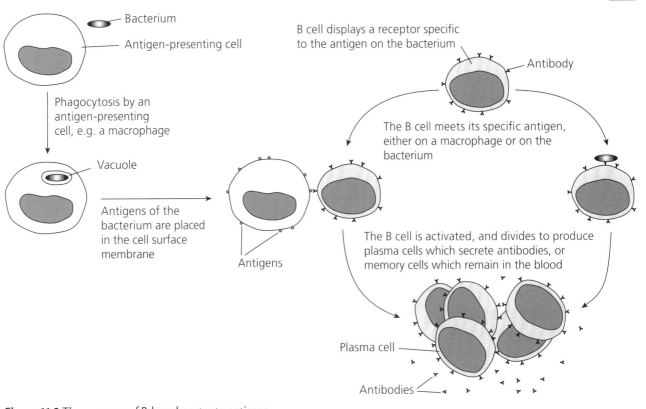

Figure 11.2 The response of B-lymphocytes to antigens

Bacterium

Antigen-presenting cell

Phagocytosis by an antigen-presenting cell, e.g. a macrophage

Vacuole

Antigens of the bacterium are placed in the cell surface membrane

Antigens

B cell displays a receptor specific to the antigen on the bacterium

Antibody

The B cell meets its specific antigen, either on a macrophage or on the bacterium

The B cell is activated, and divides to produce plasma cells which secrete antibodies, or memory cells which remain in the blood

Plasma cell

Antibodies

Action of T-lymphocytes

T-lymphocytes include **T helper cells** and **T killer cells**. Both of these types of cell place their specific receptors in their cell surface membranes. On encountering the relevant antigen, they are activated and divide by mitosis to form a clone.

Activated T helper cells secrete chemicals called **cytokines**. These stimulate B-lymphocytes to produce plasma cells, and stimulate monocytes and macrophages to attack and destroy pathogens.

Activated T killer cells attach to body cells that display the antigen matching their receptor. This happens when a virus invades a body cell. The T killer cell destroys the infected body cell (Figure 11.3).

Some of the clone of T cells become memory cells, which remain in the body and can react swiftly if the same pathogen invades again.

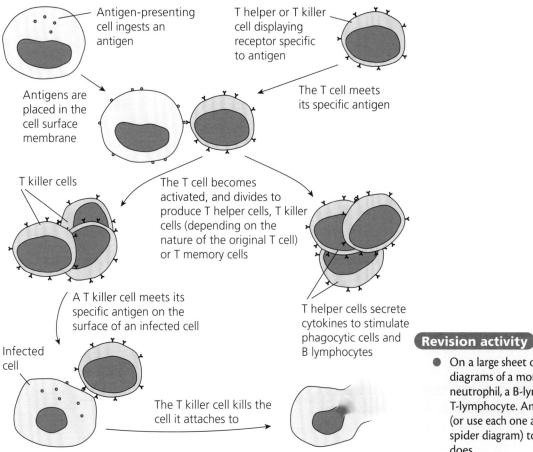

Figure 11.3 The response of T-lymphocytes to antigens

Revision activity

- On a large sheet of paper, make diagrams of a monocyte, a neutrophil, a B-lymphocyte and a T-lymphocyte. Annotate each one (or use each one as the centre of a spider diagram) to explain what it does..

Antibodies and vaccination

Antibodies ———————————————— Revised

Antibodies are glycoproteins called **immunoglobulins** that are secreted by plasma cells in response to the presence of antigens (Figure 11.4).

The variable region of the immunoglobulin molecule is specific to the particular clone of B-lymphocytes that secreted it. It is able to bind with a particular type of antigen molecule. Immunoglobulins can:

- stick bacteria together, making it impossible for them to divide or making it easier for phagocytes to destroy them

Typical mistake

Do not confuse antibodies with antibiotics (see p. 68).

- neutralise toxins (poisonous chemicals) produced by pathogens
- prevent bacteria from sticking to body tissues
- bind to viruses and prevent them infecting cells

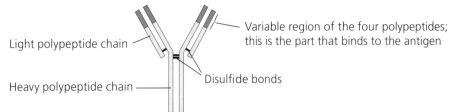

Light polypeptide chain

Variable region of the four polypeptides; this is the part that binds to the antigen

Heavy polypeptide chain

Disulfide bonds

Figure 11.4 An immunoglobulin molecule

Immunity

Revised

A person is immune to a disease if the pathogen that causes the disease is unable to reproduce in the body and cause illness. This happens when the body already contains, or is able rapidly to make, large quantities of antibodies against the antigens associated with the pathogen.

Active immunity occurs when the person's own lymphocytes make the antibody. This could be **natural**, as a result of the person having previously had the disease and forming B or T memory cells. It could also be **artificial**, as a result of **vaccination**. This involves introducing weakened pathogens into the body. The lymphocytes react to the antigens on the pathogens by producing antibodies and memory cells.

Passive immunity occurs when antibodies from elsewhere are introduced into the body. In a young baby this can be **natural**, as the baby acquires antibodies from its mother in breast milk. It can also be **artificial**, as the result of an injection of antibodies obtained from another animal.

Active immunity lasts much longer than passive immunity, because memory cells last a long time, whereas individual antibodies do not. Injections of antibodies, however, can be useful if a person requires instant immunity, for example if an aid worker is about to travel to an environment where risk of a disease such as hepatitis is high.

Revision activity

- Construct a table to compare the features of natural active immunity, artificial active immunity, natural passive immunity and artificial passive immunity. Your table should include how they are acquired, what is happening in the body, and how long the immunity lasts.

Global eradication of infectious disease

Revised

The World Health Organization has helped to organise worldwide campaigns to eliminate the serious infectious diseases smallpox and poliomyelitis.

Smallpox has been successfully eradicated by vaccinating large numbers of children with weakened viruses similar to those that cause smallpox. This succeeded because the vaccine is highly effective. The programme involved the vaccination of all relatives and contacts of anyone who had the disease — this is called ring vaccination. The virus did not mutate, so the same vaccine could be used everywhere.

Diseases that have *not* been successfully eradicated include the following:
- **Measles**. This is partly because several successive doses of vaccine are required to produce immunity, especially in children who have weak immune systems owing to poor diet or living conditions. The virus is highly infectious, so a very high percentage of people must be vaccinated to ensure a population is free of it. Booster vaccinations are also needed. This is difficult to achieve in places where infrastructure is poor.
- **TB**. The BCG vaccination gives only partial immunity, although new vaccines are now being developed, which it is hoped will be more effective. TB is

difficult to treat because the bacteria live inside body cells. Many strains have developed resistance to antibiotics.

- **Malaria**. No effective vaccine has yet been developed against *Plasmodium*. This is a eukaryotic organism, not a bacterium or virus, and is not affected by T-lymphocytes or by antibodies produced by B-lymphocytes.
- **Cholera**. This disease is caused by the bacterium *Vibrio cholerae*. In the body, it lives and reproduces in the intestine, which is outside the body tissues and not easily reachable by lymphocytes or antibodies. Current cholera vaccines are ineffective, partly because injected vaccines do not readily reach the intestines. Oral vaccines are being developed, which are proving more effective.

Vaccines are, of course, completely ineffective against any diseases that are not caused by pathogens, such as sickle cell anaemia.

Monoclonal antibodies

Revised

Monoclonal antibodies are large quantities of identical antibodies produced by a clone of genetically identical plasma cells. They have many uses in medicine (see below).

Producing monoclonal antibodies

Plasma cells are not able to divide, so it is not possible to make a clone of them. A plasma cell is therefore fused with a cancer cell, to produce a **hybridoma** cell. This cell retains the ability of the plasma cell to secrete a particular antibody, and also the ability of the cancer cell to divide repeatedly to produce a clone (Figure 11.5).

1. A mouse or other organism is injected with an antigen that will stimulate the production of the desired antibody.
2. B-lymphocytes in the mouse that recognise this antigen divide to form a clone of plasma cells able to secrete an antibody against it.
3. B-lymphocytes are then taken from the spleen of the mouse. These are fused with cancer cells to produce hybridoma cells.
4. To identify the hybridoma cells that produce the desired antibody, individual cells are separated out. Each hybridoma cell is tested to see if it produces antibody on exposure to the antigen. Any that do are cultured so that they produce a clone of cells all secreting the desired antibody.

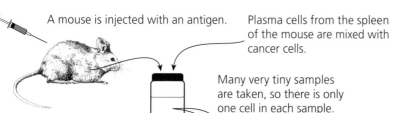

A mouse is injected with an antigen.

Plasma cells from the spleen of the mouse are mixed with cancer cells.

Many very tiny samples are taken, so there is only one cell in each sample.

Every sample is tested to find a hybridoma cell – the product of the fusion of a mouse plasma cell producing antibody to the antigen and a cancer cell.

The sample containing the hybridoma cell is cloned. Being a cancer cell, it grows in unlimited quantities in the laboratory.

Figure 11.5 How monoclonal antibodies are produced

Uses of monoclonal antibodies

Monoclonal antibodies can be used in the diagnosis of disease, and in pregnancy testing.

Their big advantage is that the tests can be carried out quickly, often producing results in minutes or hours.

Diagnosing disease

Monoclonal antibodies can be used in an ELISA test to diagnose an infectious disease, by detecting the presence of particular antigens in the blood (Figure 11.6). ELISA stands for enzyme-linked immunosorbent assay.

1 The monoclonal antibody is immobilised on the surface of a small container, such as a small glass well. The liquid to be tested (for example, blood plasma) is added to the well. If the antigens are present, they will bind to the antibodies.

2 The contents of the well are then rinsed out. The antigens stay tightly bound to the antibodies.

3 Now another solution containing the same monoclonal antibodies is added to the well. These antibodies also have a reporter enzyme combined with them. They bind with the antigens already attached to the antibodies in the well. The well is again rinsed out, so the enzymes will all be washed away unless they have bound with the antigens.

4 The substrate of the enzyme is then added. If the enzyme is present — which is only the case if the antigen being investigated was present — then the substrate is changed to a coloured substance by the enzyme.

The colour change therefore indicates the presence of the antigen in the fluid being tested.

Figure 11.6 How an ELISA test works

Pregnancy testing

The urine of a pregnant woman contains the hormone human chorionic gonadotrophin, hCG. Monoclonal antibodies can be used to detect its presence in urine.

There are many different types of pregnancy testing kits. One type uses a plastic strip containing three bands:

● The R band contains immobilised monoclonal antibodies that can bind with hCG. These antibodies have been combined with an enzyme.

● The T band contains more antibodies that can bind with hCG, and also coloured substrates for the enzymes in the R band.

● The C band is used to check that the strip is working.

When the end of the strip is dipped into urine, the urine moves up the strip by capillary action. If it contains hCG, this binds with the monoclonal antibodies in the R band. The complexes of hCG, antibodies and enzymes break free from the strip, and continue moving up it as the urine seeps upwards.

When they reach the T band, the enzymes attached to the antibodies cause the substrate to react, producing a coloured product. This produces a coloured stripe on the test strip.

Treating disease

Monoclonal antibodies can be produced that will bind with particular proteins on the surface of body cells. For example, the monoclonal antibody rituximab binds with a protein called CD20, which is found only on B lymphocytes. This can be useful in the treatment of a type of cancer called non-Hodgkin lymphoma, in which B cells divide uncontrollably. When rituximab binds to the B cells, it makes them 'visible' to the immune system, which destroys them. New B cells are made in the bone marrow, and these replacement cells may not be cancerous.

AS experimental skills and investigations

Skills and mark allocations

Almost one quarter of the total marks for your AS examination are for experimental skills and investigations. These are assessed through a practical examination.

There is a total of 40 marks available for the practical examination. Although the questions are different each time, the number of marks assigned to each skill is always the same. This is shown in the table below.

Skill	Total marks	Breakdown of marks	
Manipulation, measurement and observation, MMO	16	Successfully collecting data and observations	8 marks
		Making decisions about measurements or observations	8 marks
Presentation of data and observations, PDO	12	Recording data and observations	4 marks
		Displaying calculations and reasoning	2 marks
		Data layout	6 marks
Analysis, conclusions and evaluation, ACE	12	Interpreting data or observations and identifying sources of error	6 marks
		Drawing conclusions	3 marks
		Suggesting improvements	3 marks

The syllabus explains each of these skills in detail, and it is important that you read the appropriate pages in the syllabus so that you know what each skill is, and what you will be tested on.

The next few pages explain what you can do to make sure you get as many marks as possible for each of these skills. They give you guidance in how you can build up your skills as you do practical work during your course, and also how to do well in the examination itself. They are not arranged in the same order as in the syllabus, or in the table above. Instead, they have been arranged by the kind of task you will be asked to do, either in practical work during your biology course or in the examination.

There is a great deal of information for you to take in, and skills for you to develop. The only way to do this really successfully is to do lots of practical work, and gradually build up your skills bit by bit. Don't worry if you don't get everything right first time. Just take note of what you can do to improve next time — you will steadily get better and better.

The practical examination questions

There are usually two questions in the practical examination. The examiners will take care to set questions that are *not exactly the same* as any you have done before. It is possible that there could be three shorter questions instead of two longer ones, so do not be surprised if that happens.

It is important that you do exactly what the question asks you to do. Students often lose marks by doing something they have already practised, rather than doing what the question actually requires.

Question 1 is likely to be what is sometimes called a 'wet practical'. For example, it could be:

- an investigation into the activity of an enzyme
- an osmosis experiment
- tests for biological molecules

This question will often ask you to investigate the effect of one factor on another — for example, the effect of enzyme concentration on rate of reaction, or the effect of leaf area on the rate of transpiration.

Question 2 is likely to involve making drawings from a specimen. This could be a real specimen, or it could be a photograph. You may be asked to use a microscope, a stage micrometer and eyepiece graticule, or images of them, to work out the magnification or size of the specimen.

The two questions are designed to take approximately equal amounts of your time. You should therefore aim to spend about 1 hour on each question.

Expert tip

During your course:
- Every time you do a practical during your AS course, time yourself. Are you working quickly enough? You will probably find that you are very slow to begin with, but as the course progresses try to work a little faster as your confidence improves.

In the exam:
- Do exactly what the question asks you to do. This is unlikely to be exactly the same as anything you have done before.
- Leave yourself enough time to do each question, spending an appropriate number of minutes on each one.

How to do well in the practical exam

Variables

Revised

Many of the experiments that you will do during your AS course, and usually Question 1 in the examination paper, will investigate the effect of one factor on another. These factors are called **variables**.

The factor that you change or select is called the **independent variable**. The factor that is affected (and that you measure when you collect your results) is the **dependent variable**. The table below shows some examples.

The **independent variable** is the factor that you change in an investigation.

The **dependent variable** is the factor that changes as the result of changes in the independent variable.

Investigation		Independent variable	Dependent variable
1	Investigation into the effect of temperature on the rate of breakdown of hydrogen peroxide by catalase	Temperature	Volume of oxygen produced per minute
2	Investigating the effect of immersion in solutions of different sucrose concentration on the change in length of potato strips	Sucrose concentration	Change in length of potato strip
3	Testing the hypothesis: the density of stomata on the lower surface of a leaf is greater than the density on the upper surface	Upper or lower surface of the leaf	Number of stomata per cm^3
4	Investigation into the effect of leaf area on transpiration rate	Total area of leaves	Rate of movement of meniscus

We will keep referring back to these four examples in the next few pages, so you might like to put a marker on this page so you can easily flip back to look at the table as you read.

If you are investigating the effect of one variable on another, then you need to be sure that there are no other variables that might be affecting the results. It is important to identify these and — if possible — keep them constant. These are sometimes called **control variables**.

Making decisions about the independent variable

You may have to make your own decisions about the range and interval of the independent variable.

Let's think about Investigation 1 in the table above — investigating the effect of temperature on the rate of breakdown of hydrogen peroxide by catalase.

The independent variable is the temperature. First, decide on the **range** of temperatures you will use. The range is the spread between the highest and lowest value. This will be affected by:

● the apparatus you have available to you, which will determine the possible range of temperatures you can produce. In this case, you will probably be using a water bath. If you are lucky, you may have a thermostatically controlled water bath, but in the exam you will probably have to use a beaker of water whose temperature you can control by adding ice or by heating it.

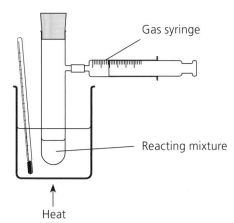
Gas syringe

Reacting mixture

Heat

Changing the independent variable

● your knowledge about the range of temperature over which the rate of activity of the enzyme is likely to be affected. Even if you could manage it, there would not be much point in trying temperatures as low as −50 °C or as high as 200 °C. However, you probably know that various enzymes can have optimum temperatures anywhere between 20 °C and 80 °C, so you should include these values in the range.

Next, decide on the **intervals** that you will use. The interval is the distance between the values that you choose. This will be affected by:

● the number of different values you can fit in within your chosen range, and how much time you have available to you. For example, you might ideally like to use intervals of 5 °C, so that you set up water baths at 0 °C, 5 °C and so on, all the way up to 80 °C. But obviously that would not be sensible if you only have five water baths, or if you only have 1 hour to do the experiment.

● the number of results you need to obtain. You are going to be looking for any pattern in the relationship between the independent variable (temperature) and the dependent variable (rate of reaction). You will need at least five readings to see any pattern. There is really no point trying to draw a graph if there will be fewer than five points on it. So, if your range of temperatures is 0 °C to 80 °C, you could use intervals of 20 °C. This would give you five readings — 0, 20, 40, 60 and 80 °C.

Producing different concentrations of a solution

In Investigation 2, Investigating the effect of immersion in solutions of different sucrose concentration on the change in length of potato strips, the independent variable is the concentration of a solution. You may be given a sucrose solution of a particular concentration, and then be asked to produce a suitable range of concentrations to carry out the experiment.

The **range** you should use will usually be from 0 (distilled water) up to the concentration of the solution you have been given (because obviously you cannot easily make that into a more concentrated solution).

The **intervals** you use could be either:

● all the same distance apart, for example concentrations of 0.8, 0.6, 0.4 and 0.2 mol dm^{-3} (and, of course, 0.0 mol dm^{-3})
● produced by using serial dilutions to make concentrations of 0.1, 0.01 and 0.001 mol dm^{-3} (and, of course, 0.0 mol dm^{-3})

How to... Produce a range of solutions of different concentrations from one given concentration

Let's say you need 10 cm^3 of sucrose solution of each concentration.

Producing a range with equal intervals

Take a particular volume of your original solution and place it in a clean tube. Add distilled water to make it up to 10 cm^3.

Then do the same again, using a different volume of the original solution.

The table below gives some examples.

Producing a range of concentrations with equal intervals from a 1.0 mol dm^{-3} solution

Volume taken of original 1 mol dm^{-3} solution/cm^3	Volume of distilled water added/cm^3	Concentration of solution produced/ mol dm^{-3}
10	0	1.0
8	2	0.8
6	4	0.6
4	6	0.4
2	8	0.2
0	10	0.0

Producing a range using serial dilutions

You could be asked to make up a series of solutions in which each one has a concentration that is one tenth of the previous one.

Take 1 cm^3 of your original solution and place it in a clean tube.

Add distilled water to make it up to 10 cm^3.

Now mix this new solution really well, and then take 1 cm^3 of it. Put this into a clean tube and make it up to 10 cm^3.

Keep doing this, each time taking 1 cm^3 from the new solution.

The table below summarises this.

Producing a range of concentrations using serial dilution of a 1.0 mol dm^{-3} solution

Solution used/ mol dm^{-3}	Volume taken of solution/cm^3	Volume of distilled water added/cm^3	Concentration of solution produced/ mol dm^{-3}
1.0	10	0	1.0
1.0	1	9	0.1
0.1	1	9	0.01
0.01	1	9	0.001
0.001	1	9	0.0001
0.0001	1	9	0.00001

You could also be asked to make up solutions where each is one half of the concentration of the previous solution.

Continuous and discontinuous variables

In Investigation 1, the independent variable (temperature) is **continuous**. This means that we can choose any value within the range we have decided to use. This is also true for Investigation 2, where we can choose any value of concentration within the range we have decided to use.

Sometimes, however, the independent variable is **discontinuous**. This means that there is only a limited number of possible values. For example, in Investigation 3, Testing the hypothesis: the density of stomata on the lower surface of a leaf is greater than the density on the upper surface, the independent variable has only two possible 'values' — either the upper surface of the leaf, or the lower surface of the leaf. So you don't have any choice about the range or intervals at all!

Controlling the control variables

In your experiment, it is important to try to make sure that the only variable that could be affecting the dependent variable is the independent variable that you are investigating. If you think there are any other variables that might affect it, then you must try to keep these constant.

Look back at the table on p. 76.

In Investigation 1, the important control variables would be the concentration and volume of the enzyme solution and the concentration and volume of the hydrogen peroxide solution. Changes in any of these would have a direct effect on the rate of reaction.

In Investigation 2, the important control variables would be the dimensions of the potato strips and the potato tuber from which they came. You also need to think about time, but here the important thing is that the strips are left in the solution for long enough for equilibrium to be reached — after that, it doesn't really matter if one is left for slightly longer than another. You also need to be sure that all the strips are completely immersed in the solution, although the actual volume of the solution doesn't matter. Temperature, too, will not affect the final result, but it could affect the speed at which equilibrium is reached — if you leave the strips for long enough, then it does not really matter if the temperature varies.

In biology, we often want to do experiments where it is not possible to control all the variables. For example, we might want to investigate the effect of body mass index on heart rate when at rest. There are all sorts of other variables that might affect resting heart rate, such as gender, age, fitness, when a person last ate and so on. In this case, we just have to do the best we can, for example, by limiting our survey to males between the ages of 20 and 25. If we can collect results from a large *random* sample among this group of males, then we can hope that at least we will be able to see if there appears to be a relationship between our independent and dependent variables.

When to measure the dependent variable

In many experiments you will need to decide when, and how often, you should take a reading, observation or measurement of the dependent variable.

- With some investigations, you will need to leave things long enough for whatever is happening to finish happening. This would be important in Investigation 2, where you would need to leave the potato strips in the sucrose solutions for long enough for equilibrium to be reached.

A **continuous variable** can have any value within its range.

A **discontinuous variable** has only a limited number of possible values.

Expert tip

During your course:
- Every time you do an experiment, identify and write down the independent variable and the dependent variable.
- Every time you do an experiment, think about the *range* and the *intervals* of the values you are using for the independent variable. For your own benefit, write down what the range is and what the intervals are, just to help you to think about them.
- Learn how to make up dilutions from a solution of a given concentration, and practise doing this until you feel really confident about it.

In the exam:
- Read the question carefully, then identify the independent variable and the dependent variable (even if the question does not ask you to do this).
- Next, decide if the independent variable is continuous or discontinuous (see above).
- If it is continuous, read the question carefully to see if you have been told the range and intervals to use, or if you are being asked to decide these for yourself.

Expert tip

During your course:
- Every time you do an experiment, think about which variables you have been told to control, or make your own decision about which ones are important to control. Get to know the standard ways of controlling variables such as temperature (use water baths), pH (use buffer solutions) and other variables.

In the exam:
- If you are not told what variables to control, then think about these carefully before deciding what you will control and how you will do it.

- With some investigations, you may need to begin taking readings straight away. This would be important in Investigation 1, where you should begin measuring the volume of oxygen released each minute from time 0, which is the moment that the enzyme and its substrate are mixed.
- With some investigations, you may need to allow time for a process to settle down to a steady rate before you begin to take readings. This would be important in Investigation 4, where you would be measuring the rate of transpiration in a particular set of conditions.

Expert tip

During your course:
- Every time you do an experiment where time is involved, think about why you should start timing from a particular moment, and when and why you should take readings.

In the exam:
- Think carefully about whether or not time is important. If you think it is, then decide when you will start taking readings, and how often you will take them. Remember that if you are going to use them to draw a graph, you will need at least five points to plot.

Taking measurements

Revised

You will often be asked to take measurements or readings. In biology, these are most likely to be length, mass, time, temperature or volume. You could be taking readings from a linear scale (for example, reading temperature on a thermometer, reading volume on a pipette, or reading length on a potometer tube). You could be reading values on a digital display, for example reading mass on a top pan balance or time on a digital timer.

There are some special terms that are used to describe measurements, and the amount of trust you can put into them. It is easiest if we think about them in terms of a particular experiment, so let's concentrate on Investigation 1. Look back at p. 76 to remind yourself what is being measured.

Validity This is about whether what you are measuring is what you actually *intend* to measure. For example, in Investigation 1, does measuring the volume of oxygen in the gas syringe each minute really tell you about the rate of reaction? It is a valid method in this instance, because the volume of oxygen given off per unit time is directly related to the rate at which the reaction is taking place.

Reliability This is how well you can trust your measurements. Reliable results are ones that are repeatable. This could be affected by various factors, such as whether you are able to take a reading at the precise time you intended to.

Accuracy An accurate reading is a true reading. For your readings of volume to be accurate, then the gas syringe must have been calibrated correctly, so that when it says the volume of gas is $8.8\,cm^3$, then there really is exactly $8.8\,cm^3$ of gas in there.

Precision If you were able to put exactly $8.8\,cm^3$ of gas into your gas syringe, and it read $8.8\,cm^3$ every time, then your readings have a high degree of precision. If, however, the syringe did not always read the same value (so there was variation in its readings, even though the actual volume of gas was exactly the same), then your measurements are less precise.

Resolution You probably already know this term, because we use it in microscopy to tell us the degree of detail that we can see. The smaller the detail, the higher the resolution. It means very much the same thing with a measuring instrument — the smaller the division on the scale of the measuring instrument, the higher its resolution. So, for example, a $10\,cm^3$ gas syringe marked off every $0.5\,cm^3$ has a higher resolution than a $20\,cm^3$ gas syringe marked off every $1\,cm^3$. If you get a choice, then go for the instrument with the highest resolution to make your measurements — so long as it can cover the range that you need.

Uncertainty in measurements — estimating errors

Whenever you take a reading or make a measurement, there will be some uncertainty about whether the value is absolutely correct. These uncertainties are **experimental errors**. Every experiment, no matter how well it has been designed, no matter how carefully it has been carried out and no matter how precise and accurate the measuring instruments, has this type of error.

You may be asked to estimate the size of the errors in your measurements. *This is nothing to do with how well you have made the measurements* — the examiners do not want to know about 'mistakes' that you might have made, such as misreading a scale or taking a reading at the wrong time. It is all about the inbuilt limitations in your measuring device and its scale.

- In general, *the size of the error is half the value of the smallest division on the scale.* For example, if you have a thermometer that is marked off in values of 1°C, then every reading that you take could be out by 0.5°C. You can show this by writing: 21.5°C ± 0.5°C.

- If your recorded result involves measuring *two* values — for example, if you have measured a starting temperature and then another temperature at the end, and have calculated the rise in temperature — then this error could have occurred for both readings. *The total error is therefore the sum of the errors for each reading.* Your final value for the change of temperature you have measured would then be written: 18.0°C ± 1.0°C.

> An **experimental error** is a factor that reduces the reliability of your results, such as limitations of the technique or apparatus used, or inability to control a variable that affects the results. It must not be confused with a human mistake.

> **Expert tip**
>
> Every time you take a reading or make a measurement, get into the habit of working out and writing down the error (uncertainty) in each reading.

Recording measurements and other data

Revised

You will often need to construct a table in which to record your measurements, readings and other observations. It is always best to design and construct your results table *before* you begin your experiment, so that you can write your readings directly into it as you take them.

Let's think about Investigation 2 again. You have made your decisions about the range and intervals of the independent variable (concentration of solution) — you have decided to use six concentrations ranging from $0.0\,mol\,dm^{-3}$ to $1.0\,mol\,dm^{-3}$. Your dependent variable is the change in length of the potato strips, and you are going to find this by measuring the initial length and final length of each strip.

These are the things you need to think about when designing your results table:
- The **independent variable** should be in the first column.
- The **readings** you take are in the next columns.
- Sometimes, these readings actually *are* your dependent variable. In this experiment, however, you are going to have to use these readings to *calculate* your dependent variable, which is the change in length of the strips. So you need to have another column for this. This comes at the end of the table. In fact, you really need to work out the *percentage* change in length of the strips, as this will allow for the inevitable variability in the initial lengths of the strips.

The table could look like this:

Results table for Investigation 2

Concentration of sucrose solution/ $mol\,dm^{-3}$	Initial length of potato strip/mm	Final length of potato strip/mm	Change in length of potato strip/mm	Percentage change in length of potato strip
0				
0.2				
0.4				
0.6				
0.8				
1.0				

Notice:

- The table has been clearly drawn, with lines separating all the different rows and columns. Always use a pencil and ruler to draw a results table.
- Each column is fully headed, including the unit in which that quantity is going to be measured. The unit is preceded by a slash (/). You can use brackets instead, for example, concentration of sucrose solution (mol dm^{-3}).
- The slash always means the same thing. It would be completely wrong to write: concentration of sucrose solution/mol/dm^3 as the heading of the first column. That would be really confusing. If you are not happy using negative indices like dm^{-3}, you can always write 'per' instead. So it would be fine to write: concentration of sucrose solution/mol per dm^3.
- The columns are all in a sensible order. The first one is the independent variable, so you can write these values in straight away, as you have already decided what they will be. The next thing you will measure is the initial length of the strip, then the final length. Then you will calculate the change in length, and finally you will calculate the percentage change in length.

So now you are ready to do your experiment and collect your results. Here is what your table might look like.

Completed results table for Investigation 2

Concentration of sucrose solution/ mol dm^{-3}	Initial length of potato strip/mm	Final length of potato strip/mm	Change in length of potato strip/mm	Percentage change in length of potato strip
0	49.5	52.0	+2.5	+5.1
0.2	50.0	52.0	+2.0	+4.0
0.4	50.5	51.5	+1.0	+2.0
0.6	50.0	50.5	+0.5	+1.0
0.8	49.0	48.0	−1.0	−2.0
1.0	49.5	48.0	−1.5	−3.0

Notice:

- All the measurements in the second two columns were made to the nearest 0.5 mm. This is because the smallest graduation on the scale on the ruler was 1 mm, so it was possible to estimate the length to the nearest 0.5 mm. (Have a look at the scale on your ruler, and you will see that this is sensible.) Even if you decide that a length is exactly 50 mm, you must write in the next decimal place for consistency, so you would write 50.0.
- The values in the 'change in length' column each show whether they were an increase or a decrease.
- The percentage change in length is calculated like this:

$$\frac{\text{change in length}}{\text{initial length}} \times 100$$

  (Do make sure you remember to take a calculator into the exam with you.)
- Each percentage change in length has been rounded up to one decimal place, for consistency with the change in length. For example, the calculation in the first row gives 6.0606, which you should round up to 6.1. The calculation in the sixth row gives 4.0404, which rounds down to 4.0.

Repeats

It is a good idea to do **repeats**. This means that, instead of getting just one reading for each value of your independent variable, you collect two or three. You can then calculate a **mean value**, which is more likely to be a 'true' value than any of the individual ones.

Let's say that you did this for the potato strip experiment. You could have used two potato strips for each sucrose concentration, then calculated the percentage change in length for each one, then finally calculated a mean percentage change.

This means adding some extra rows and an extra column to the results table, like this:

Completed results table (with repeats) for Investigation 2

Concentration of sucrose solution/ mol dm^{-3}	Initial length of potato strip/mm	Final length of potato strip/mm	Change in length of potato strip/ mm	Percentage change in length of potato strip	Mean percentage change in length of potato strip
0	49.5	52.0	+2.5	+5.1	+4.6
	49.0	51.0	+2.0	+4.1	
0.2	50.0	52.0	+2.0	+4.0	+4.0
	50.5	52.5	+2.0	+4.0	
0.4	50.5	51.5	+1.0	+2.0	+2.5
	49.5	51.0	+1.5	+3.0	
0.6	50.0	50.5	+0.5	+1.0	+0.5
	51.0	51.0	0.0	0.0	
0.8	49.0	48.0	−1.0	−2.0	−2.0
	50.5	49.5	−1.0	−2.0	
1.0	49.5	48.0	−1.5	−3.0	−3.0
	50.5	49.0	−1.5	−3.0	

Notice:
- All of this information has been put in a single results table. This makes it much easier for someone to read and find all the information they need.
- The numbers in the final column have again been rounded up to one decimal place.

Qualitative observations

The results table for the potato strip experiment contains numerical values — they are **quantitative**. Sometimes, though, you may want to write descriptions in your results table, for example a colour that you observed. These are **qualitative** observations. If you are recording colours, write down the actual colour — do not just write 'no change'.

Use simple language that everyone can easily understand. Avoid using terms that are difficult for the examiner to interpret, such as 'yellowish-green'. Think about what is important — perhaps it is that *this* tube is a darker or lighter green than *that* tube. Using simple language such as 'dark green' or 'a lighter green than tube 1' is fine.

Graphs and other ways of displaying data

Revised

When you have collected your data and completed your results table, you will generally want to display the data so that anyone looking at them can see any patterns.

Line graphs

Line graphs are used when both the independent variable and the dependent variable are continuous (see p. 79). This is the case for the potato strip data above. The graph can help you to decide if there is a relationship between the independent variable and the dependent variable. This is what a line graph of these data might look like.

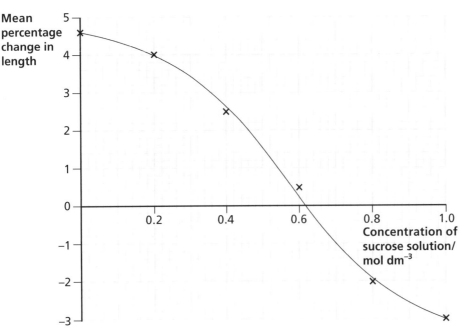

Graph of the results of Investigation 2

Notice:

- The independent variable goes on the *x*-axis, and the dependent variable goes on the *y*-axis.
- Each axis is fully labelled with units. You can just copy the headings from the appropriate columns of your results table.
- The scales on each axis should start at or just below your lowest reading, and go up to or just above your highest reading. Think carefully about whether you need to begin at 0 on either of the axes, or if there is no real reason to do this.
- The scales use as much of the width and height of the graph paper as possible. If you are given a graph grid on the exam paper, the examiners will have worked out a sensible size for it, so you should find your scales will fit comfortably. The greater the width and height you use, the easier it is to see any patterns in your data once you have plotted them.
- The scale on each axis goes up in regular steps. Choose something sensible, such as 1s, 2s, 5s or 10s. If you choose anything else, such as 3s, it is practically impossible to read off any intermediate values. Imagine trying to decide where 7.1 is on a scale going up in 3s.
- Each point is plotted very carefully with a neat cross. Do not use just a dot, as this may not be visible once you have drawn the line. You could, though, use a dot with a circle round it.
- A smooth best-fit line has been drawn. This is what biologists do when they have good reason to believe there is a smooth relationship between the independent and dependent variables. You know that your individual points may be a bit off this line (and the fact that the two repeats for each concentration were not always the same strongly supports this view), so you can actually have more faith in there being a smooth relationship than you do in your plots for each point.

Sometimes in biology (it does not often happen in physics or chemistry!) you might have more trust in your individual points than in any possible smooth relationship between them. If that is the case, then you do not draw a best-fit curve. Instead, join the points with a very carefully drawn straight line, like this:

Expert tip

During your course:
- Get plenty of practice in drawing graphs, so that it becomes second nature always to choose the correct axes, to label them fully and to choose appropriate scales.

In the exam:
- Take time to draw your graph axes and scales — you may need to try out two or even three different scales before finding the best one.
- Take time to plot the points — and then go back and check them.
- Use a sharp HB pencil to draw the line, taking great care to touch the centre of each cross if you are joining points with straight lines. If you go wrong, rub the line out completely before starting again.
- If you need to draw two lines on your graph, make sure you label each one clearly.

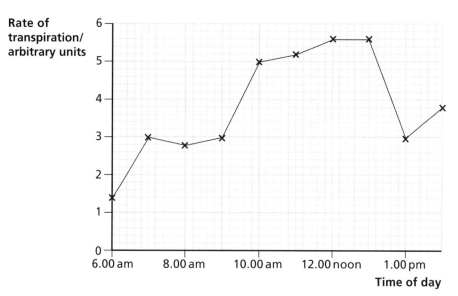

Graph where we are not sure of the pattern in the relationship between the independent and dependent variables

You may be asked to read off an intermediate value from the graph you have drawn. It is always a good idea to use a ruler to do this — place it vertically to read a value on the x-axis, and horizontally to do the same on the y-axis. You can draw in faint vertical and horizontal pencil lines along the ruler. This will help you to read the value accurately.

You could also be asked to work out the gradient of a line on a graph. This is explained on p. 28.

(This is explained on p. 28.)

Histograms

A histogram is a graph where there is a continuous variable on the x-axis, and a frequency on the y-axis. For example, you might have measured the length of 20 leaves taken from a tree. You could plot the data like this:

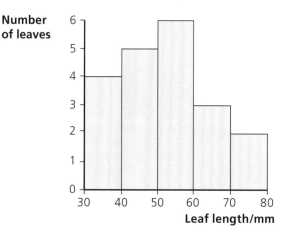

A frequency histogram

Notice:

● The numbers on the x-axis scale are written *on* the lines. The first bar therefore includes all the leaves with a length between 30 and 39 mm. The next bar includes all the leaves with a length between 40 and 49 mm, and so on.
● The bars are all the same width.
● The bars are all touching — this is important, because the x-axis scale is continuous, without any gaps in it.

> **Expert tip**
>
> During your course:
> ● Make sure you know how to read off an intermediate value from a graph accurately, and how to calculate a gradient.
>
> In the exam:
> ● Take time over finding intermediate values on a graph — if you rush it is very easy to read off a value that is not quite correct.

Bar charts

A bar chart is a graph where the independent variable is made up of a number of different, discrete categories and the dependent variable is continuous. For example, the independent variable could be type of fruit juice, and the dependent variable could be the concentration of glucose in the juice.

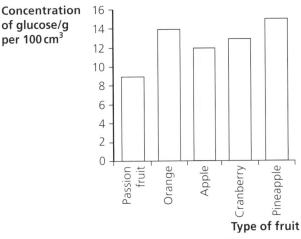

Bar chart showing concentration of glucose in different types of fruit juice

Notice:

- The x-axis has an overall heading (type of fruit), and then each bar also has its own heading (orange, apple and so on on).
- The y-axis has a normal scale just as you would use on a line graph.
- The bars are all the same width.
- The bars do not touch.

Drawing conclusions and interpreting your data
Revised

Once you have collected, tabulated and displayed your results, you can use them to draw a conclusion.

When you are thinking about a conclusion, look right back to the start of your experiment where you were told (or you decided) what you were to investigate. For example:

- In Investigation 1, Investigating the effect of temperature on the rate of breakdown of hydrogen peroxide by catalase, your conclusion should provide a statement about the relationship between the two variables.
- In Investigation 2, Investigating the effect of immersion in solutions of different sucrose concentration on the change in length of potato strips, your conclusion should state the relationship between the concentration of sucrose solution and the change in length of the potato strips.
- In Investigation 3, Testing the hypothesis: the density of stomata on the lower surface of a leaf is greater than the density on the upper surface, your conclusion should say whether your results support or disprove this hypothesis.

Explaining your reasoning

There will often be marks for explaining how you have reached your conclusion. Your reasoning should refer clearly to your results. For example, your conclusion to Investigation 2 (whose results are shown in the table on p. 83) might be as follows:

- A sucrose solution with a concentration of $0.6\,mol\,dm^{-3}$ and below caused an increase in length of the potato strips. A sucrose solution with a concentration of $0.8\,mol\,dm^{-3}$ and above caused a decrease in length of the potato strips. From the graph, the solution that I would expect to cause no change in length of the strips would be 0.62 mol dm^{-3}.

- The strips gained in length because they took up water, which was because the water potential of the sucrose solution was greater than the water potential in the potato cells. This therefore means that the water potential inside the potato cells was the same as the water potential of a 0.62 mol dm^{-3} sucrose solution.

Showing your working, and significant figures

You may be asked to carry out a calculation and to show your working. There will be marks for doing this. If you do not show your working clearly, you will not get full marks, even if your answer is absolutely correct.

For example, imagine you have measured four lengths as 46 mm, 53 mm, 52 mm and 48 mm. You are asked to calculate the mean and to show your working. You should write this down properly as:

$$\text{mean length} = \frac{46 + 53 + 52 + 48}{4} \text{ mm} = 50 \text{ mm}$$

You have already seen, on p. 82, that the final answer to a calculation should have the same number of significant figures as the original numbers you were working from. If you do the calculation above, you will find the answer you get is 49.75. But the original measurements were only to two significant figures (a whole number of mm) so that is how you should give the final answer to your calculation. You must round the answer up or down to give the same number of significant figures as the original values from which you are working.

There is another example of showing your working on p. 90.

Expert tip

During your course:
- Get into the habit of describing the main steps in your reasoning when drawing a conclusion, and using evidence from the results to support it.
- Get into the habit of taking time to set out all your calculations very clearly, showing each step in the process.
- Get into the habit of giving final numerical answers to calculations to the same number of significant figures as the readings you took, or the values you were given.

In the exam:
- Even if you feel rushed, take time to write down the steps in calculations and reasoning fully.

Identifying sources of error

Revised

It is worth repeating that it is very important to understand the difference between experimental errors and 'mistakes'. A mistake is something that you do incorrectly, such as misreading the scale on a thermometer, or taking a reading at the wrong time, or not emptying a graduated pipette fully. Do *not* refer to these types of mistake when you are asked to comment on experimental errors.

You have already seen, on p. 81, that every measuring instrument has its own built-in degree of uncertainty in the values you read from it. You may remember that, in general, the size of the error is half the value of the smallest division on the scale.

Errors can also occur if there were uncontrolled variables affecting your results. For example, if you were doing an investigation into the effect of leaf area on the rate of transpiration, and the temperature in the laboratory increased while you were doing your experiment, then you cannot be sure that all the differences in rate of transpiration were entirely due to differences in leaf area.

Systematic and random errors

Systematic errors are ones that are the same throughout your investigation, such as intrinsic errors in the measuring instruments you are using.

Random errors are ones that can differ throughout your investigation. For example, you might be doing an osmosis investigation using potato strips taken from different parts of a potato, where perhaps the cells in some parts had a higher water potential than in others. Or perhaps the temperature in the room is fluctuating up and down.

Spotting the important sources of error

You should be able to distinguish between significant errors and insignificant ones. For example, a change in room temperature could have a significant effect on the rate of transpiration (Investigation 4) but it would not have any effect at all on the number of stomata on the upper and lower surface of a leaf (Investigation 3).

Another thing to consider is how well a variable has been controlled. If you were doing an enzyme investigation using a water bath to control temperature, then you should try to be realistic in estimating how much the temperature might have varied by. If you were using a high-quality, electronically controlled water bath, then it probably did not vary much, but if you were using a beaker and Bunsen burner then it is likely that temperature variations could indeed be significant.

Suggesting improvements

You might be asked to suggest how the investigation you have just done, or an investigation that has been described, could be improved. Your improvements should be aimed at getting more valid or reliable results to the question that the investigation was trying to answer — do not suggest improvements that would mean you would now be trying to answer a different question. For example, if you were doing an investigation to investigate the effect of *leaf area* on the rate of transpiration, do not suggest doing something to find out the effect of the *wind speed* on the rate of transpiration.

The improvements you suggest could include controlling certain variables that were not controlled, or controlling them more effectively. For example, you might suggest that the investigation could be improved by controlling temperature. To earn a mark, you must also say *how* you would control it, for example by placing sets of test tubes in a thermostatically controlled water bath.

You could also suggest using better methods of measurement. For example, you might suggest using a colorimeter to measure depth of colour, rather than using your eyes and a colour scale.

It is almost always a good idea to do several repeats in your investigation and then calculate a mean of your results. For example, if you are measuring the effect of light intensity on the rate of transpiration, then you could take three sets of readings for the volume of water taken up by your leafy shoot in 1 minute at a particular light intensity. The mean of these results is more likely to give you the true value of the rate of transpiration than any one individual result.

Expert tip

During your course:
- Every time you do an investigation, work out and write down the uncertainty in all the types of measurement that you make.
- Every time you do an investigation, think carefully about any errors that may be due to lack of control of variables — which ones might genuinely be significant?

In the exam:
- If you are asked about an investigation that seems familiar, it is tempting just to try to recall what the main errors were in the investigation that you did before. This is not a good idea, because the investigation in the exam may not be quite the same. Always think about the actual investigation in the examination question, and *think through* what the significant sources of error are.

Expert tip

During your course:
- If time allows, try to do at least two (and possibly three) repeats when you do an investigation.
- As you do an investigation, be thinking all the time about how reliable or accurate your measurements and readings are. Think about what you would like to be able to do to improve their reliability or accuracy.

In the exam:
- Be very precise in suggesting how you could improve the investigation — for example, do not just say you would control a particular variable, but say *how* you would control it.

Drawings

One of the questions in the exam is likely to involve drawing a specimen on a slide, using a microscope, or drawing from a photomicrograph (a photograph taken through a microscope).

Making decisions about what to draw

You might have to decide which part of a micrograph to draw. For example, there might be a micrograph of a leaf epidermis, and you are asked to draw two guard cells and four epidermal cells. It is really important that you do exactly as you are asked and choose an appropriate part of the micrograph.

Producing a good drawing

It is very important that you draw what you can see, not what you think you ought to see. For example, during your AS course you may have drawn a TS of a stem where the vascular bundles were arranged in a particular way, or were a particular shape. In the exam, you could be asked to draw a completely different type of vascular bundle that you have never seen before. Look very carefully and draw what you can see.

Your drawing should:
- be large and drawn using a sharp pencil (preferably HB, which can be easily erased if necessary) with no shading, using single, clear lines
- show the structure or structures in the correct proportions. The examiners will check that the overall shape and proportions of your drawing match those of the specimen. Don't worry — you do not need to be a wonderful artist — a simple, clear drawing is all that is required.
- show only the outlines of tissues if you are asked to draw a low power plan (LPP). A LPP should *not* show any individual cells. However, if you are using a microscope, you may need to go up to high power to check exactly where the edges of the tissues are.

You may be asked to label your drawing. In that case:
- use a pencil to draw label lines to the appropriate structure using a ruler, ensuring that the end of the label line actually touches the structure you are labelling
- make sure that none of your label lines cross each other
- write the labels horizontally
- write the labels outside the drawing itself

Calculating magnification or size

The use of a stage micrometer and eyepiece graticule is described on p. 9. You might be asked to do this on Paper 3.

You could also be given the magnification of an image, and asked to calculate the real size of something in the image. Below is an example of the kind of thing you might be asked to do.

This micrograph shows some cells from a moss. Notice that the magnification is given.

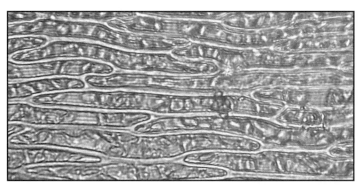

Magnification ×300

Let us say you are asked to find the mean width of a cell from the tissue in the micrograph. There are several steps you need to work through here.
- First, decide *how many* cells you are going to measure. It is generally sensible to measure a randomly selected sample of five to ten cells.
- Next, decide *which ones* you will measure. Choose cells where you can see the edges as clearly as possible, and where you can see the whole cell. If cells are evenly distributed, it is best to measure the total width of five cells in a row. That means you have to make fewer measurements, do fewer calculations and — better still — it reduces the size of the uncertainty in your measurements. However, if cells are irregularly shaped or distributed, you should measure each one individually.
- Once you have decided which five cells to measure, mark this clearly on the micrograph. It does not matter exactly how you do this — perhaps you could carefully use a ruler to draw a line across the five cells, beginning and ending exactly at the first edge of the first cell, and the last edge of the fifth cell.

During your course:
- Make sure you are familiar with the appearance of all of the structures listed in the syllabus that you could be asked about on the practical paper. You need to know the names and distribution of the tissues. Look in particular at the learning outcomes marked with [PA] at the beginning.
- Practise drawing specimens from micrographs, getting used to using your own eyes to see what is really there, rather than what you think ought to be there.
- Practise using an eyepiece graticule to help you work out the relative proportions of different parts that you are drawing.
- Take every opportunity to practise drawing specimens from micrographs or microscope slides, and either mark them yourself using a CIE-style mark scheme, or get your teacher to mark them for you. Find out what you need to do to improve, and keep working at it until you feel really confident.

In the exam:
- Take one or two sharp HB pencils, a pencil sharpener, a clean ruler that measures in mm and a good eraser.
- Settle down and take time to get your drawing of the specimen right.
- Use your eyes first, then your memory.

- Now measure the length of the line in mm and write it down.
- Next, calculate the mean length of one cell. Show clearly how you did this.
- Next, convert this length in mm to a length in μm. (Alternatively, you could do this right at the end of the calculation.)
- Next, use the magnification you have been given to convert this mean length of the image to a mean real length. Here is what your answer might look like:

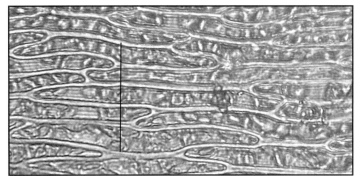

total width of 5 cells on micrograph = 29 mm

Therefore,

mean width = $\frac{29}{5}$ mm = 6 mm

$$= 6 \times 1000 = 6000\,\mu m$$

magnification = × 300

Therefore, real mean width of a cell = $\frac{6000}{300}$ μm = 20 μm

Making comparisons

Revised

You might be asked to compare the appearance of two biological specimens or structures. You could be observing these using the naked eye or a lens, or using a microscope, or you could be looking at two micrographs.

The best way to set out a comparison is to use a table. It will generally have three columns, one for the feature to be compared, and then one for each of the specimens.

For example, you might be asked to observe two leaves and record the differences between them. Your table and the first three differences might look like this:

Feature	Leaf A	Leaf B
Leaf margin	Smooth	Toothed
Veins	Parallel to each other	A central vein with branches coming off it, forming a network
Shape	Length is more than twice the maximum width	Length is less than twice the maximum width

Notice:

- The table has been drawn with ruled lines separating the columns and rows.
- The descriptions of a particular feature for each specimen are opposite one another (that is, they are in the same row).
- Each description says something positive. For example, in the first row, it would not be good to write 'not toothed' for Leaf A, as that does not tell us anything positive about the leaf margin.

Note that the practical examination is likely to ask you to describe or compare observable features, *not* functions. Do not waste time describing functions when this is not asked for.

AS exam-style questions and answers

This practice paper comprises structured questions similar to those you will meet in the exam.

You have 1 hour and 15 minutes to do the paper. There are 60 marks on the paper, so you can spend just over 1 minute per mark. If you find you are spending too long on one question, then move on to another that you can answer more quickly. If you have time at the end, then come back to the difficult one.

Some of the questions require you to recall information that you have learned. Be guided by the number of marks awarded to suggest how much detail you should give in your answer. The more marks there are, the more information you need to give.

Some of the questions require you to use your knowledge and understanding in new situations. If you come across something you have not seen before, just think carefully about it and find something that you do know that will help you to answer it.

The best answers are short and relevant — if you target your answer well, you can get a lot of marks for a very small amount of writing. Don't say the same thing several times over, or wander off into answers that have nothing to do with the question. As a general rule, there will be twice as many answer lines as marks. So you should try to answer a 3-mark question in no more than 6 lines of writing. If you are writing much more than that, you almost certainly have not focused your answer tightly enough.

Look carefully at exactly what each question wants you to do. For example, if it asks you to 'Explain', then you need to say *how* or *why* something happens, not just *describe* what happens. Many students lose large numbers of marks by not reading the question carefully.

Following each question in this practice paper, there is an answer that might get a C or D grade, followed by expert comments (shown by the icon ⊜). Then there is an answer that might get an A or B grade, again followed by expert comments. Try answering the questions yourself before looking at these.

Notice that there are sometimes more ticks on the answers than the number of marks awarded. This could be because you need two correct responses for 1 mark (e.g. Q1 (a) (i)) or because there are more potential mark points than the total number of marks available (e.g. Q1 (a) (ii)). Even if you get four or five ticks for a 3-mark question, you cannot get more than the maximum 3 marks.

Exemplar paper

Question 1

(a) The diagram shows a small part of a cell, as seen using an electron microscope.

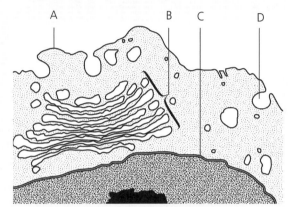

(i) Name the parts labelled A to D.
(2 marks)

(ii) Describe how part B is involved in the formation of extracellular enzymes.
(3 marks)

(b) Give two reasons, other than the presence of part B, why the cell in the diagram cannot be a prokaryotic cell. **(2 marks)**

(Total: 7 marks)

Student A

(a) (i)   A plasma membrane ✓
          B Golgi ✓
          C nucleus ✗
          D phagocyte ✗

⊜ *C is the nuclear envelope (or membrane), not the nucleus itself. A phagocyte is a cell — perhaps the student is thinking of a phagocytic vesicle. Mark: 1/2*

Student A

(ii) First, the enzymes are made by protein synthesis on the ribosomes. Then they go into the endoplasmic reticulum. Then they are taken ✓ to the Golgi where they are packaged. Then they go in vesicles ✓ to the cell membrane where they are sent out by exocytosis.

 This student has not really thought about exactly what the question was asking, and has wasted time writing about events that take place before and after the involvement of the Golgi body. There is, however, a mark for the idea that the Golgi body receives proteins that have been in the RER, and another for packaging them into vesicles. Mark: 2/3

Student A

(b) It has a nucleus ✓. And it has Golgi body ✗.

The Golgi body is part B, and this has been excluded by the question. Mark: 1/2

Student B

(a) (i)   A   cell surface membrane ✓
         B   Golgi body ✓
         C   nuclear envelope ✓
         D   vesicle ✓

All correct. Mark: 2/2

Student B

(ii) Proteins made in the RER are transported to the convex face ✓ of the Golgi body in vesicles. The vesicles fuse ✓ with the Golgi and the proteins inside are modified ✓ by adding sugars to make glycoproteins ✓. They are packaged inside membranes ✓ and sent to the cell membrane.

All correct. Mark: 3/3

Student B

(b) If it was a prokaryotic cell it would not have a nucleus ✓ and it would have a cell wall ✓.

Correct. Mark: 2/2

The diagram shows the bacterium *Mycobacterium tuberculosis,* which causes tuberculosis (TB).

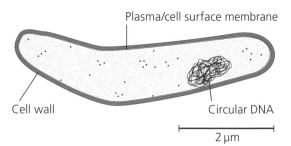

(a) *M. tuberculosis* is taken up by macrophages and multiplies inside them. Suggest how this strategy could help to protect *M. tuberculosis* from the immune response by B cells. (3 marks)

(b) In an experiment to investigate how *M. tuberculosis* avoids destruction by macrophages, bacteria were added to a culture of macrophages obtained from the alveoli of mice. At the same time, a quantity of small glass beads, equivalent in size to the bacteria, were added to the culture. The experiment was repeated using increasing quantities of bacteria and glass beads.
After 4 hours, the macrophages were sampled to find out how many had taken up either glass beads or bacteria. The results are shown in the graph. The *x*-axis shows the initial ratio of bacteria or glass beads to macrophages in the mixture. Discuss what these results suggest about the ability of macrophages to take up *M. tuberculosis*. (3 marks)

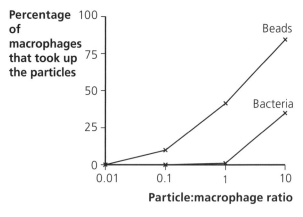

(c) When *M. tuberculosis* is present inside a phagosome of a macrophage, it secretes glycolipids that accumulate in lysosomes and prevent the lysosomes fusing with the phagosome.
Explain how this prevents the macrophage from destroying the bacterium. **(3 marks)**

(Total: 9 marks)

Student A

(a) It stops the B cells seeing them, so they do not make antibodies ✓ against them.

This is not a clear answer. B cells do not 'see', so this is not a good term to use. The 'they' in the second half of the sentence could refer to either B cells or the bacteria. Mark: 1/3

Student A

(b) The macrophages took up more glass beads than bacteria ✓. So they are not very good at taking up the bacteria ✓.

Just enough for 2 marks, although the second sentence is weak. Mark: 2/3

Student A

(c) Lysosomes contain digestive enzymes ✓, so if they do not fuse with the phagosome the bacteria will not get digested ✓.

Once again, this is the right idea, but not enough biological detail is given. Mark: 2/3

Student B

(a) B cells only become active when they meet the specific antigen ✓ to which they are able to respond. If the bacteria are inside a macrophage, then the B cells' receptors will not meet the antigen ✓ on the bacteria. This means that the B cells will not divide to produce plasma cells ✓, and will not secrete antibodies ✓ against the bacteria.

This is a good answer. Mark: 3/3

Student B

(b) The cells only started to take up any bacteria when the particle:macrophage ratio was 1 ✓. On the other hand, they took up glass beads even when the ratio was above 0.01 ✓. When the ratio of particles to macrophages was 10, only about 30% ✓ of the macrophages had taken up bacteria, whereas over 75% of them had taken up glass beads ✓. This shows the macrophages do take up the bacteria, but not as well as they take up glass beads ✓.

This is a good answer, which does attempt to 'discuss' by providing statements about the relatively low ability of the macrophages to take up the bacteria, but also stating that they do take them up. In general, it is always a good idea to quote data where they are relevant in your answer. Mark: 3/3

Student B

(c) Normally, lysosomes fuse with phagosomes and release hydrolytic enzymes ✓ into them. These enzymes then hydrolyse (digest) whatever is in the phagosome ✓. If this does not happen, then the bacteria can live inside the phagosome ✓ without being digested.

All correct. Mark: 3/3

Question 3

(a) The diagrams show a cell in various stages of the mitotic cell cycle. Name the stage represented by each diagram, and arrange them in the correct sequence. **(3 marks)**

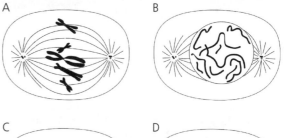

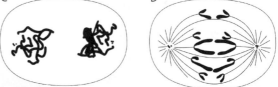

(b) Describe the role of spindle microtubules in mitosis. **(3 marks)**

(c) The graph overleaf shows the changes in the mass of DNA per cell and total cell mass during two cell cycles. Different vertical scales are used for the two lines.

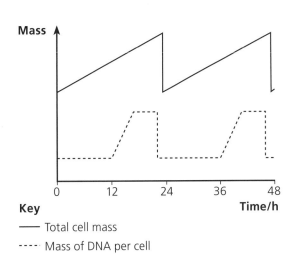

Key

— Total cell mass

---- Mass of DNA per cell

(i) **On the graph, write the letter D to indicate a time at which DNA replication is taking place.** (1 mark)

(ii) **On the graph, write the letter C to indicate a time at which cytokinesis is taking place.** (1 mark)

(d) **Describe the roles of mitosis in living organisms.** (3 marks)

(Total: 11 marks)

Student A

(a) A metaphase ✓, B prophase ✓, C telophase ✓, D anaphase ✓

Each stage is named correctly, but they are not arranged in the correct order. Mark: 2/3

Student A

(b) The spindle microtubules pull the chromatids to opposite ends of the cell ✓.

This is correct, but there is not enough here for 3 marks. Mark: 1/3

Student A

(c) Mass

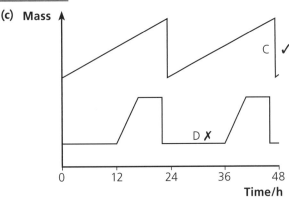

Cytokinesis is identified correctly, but DNA replication is not. The student has written D before the DNA has replicated. Mark: 1/2

Student A

(d) Mitosis is used in growth and repair. ✓

This is correct, but not a good enough answer for more than 1 mark at AS. Mark: 1/3

Student B

(a) B prophase ✓, A metaphase ✓, D anaphase ✓, C telophase ✓✓

All identified correctly, and in the right order. Mark: 3/3

Student B

(b) Spindle microtubules are made by the centrioles. They latch on to the centromeres ✓ of the chromosomes and help them line up on the equator ✓. Then they pull ✓ on the centromeres so they come apart and they pull the chromatids ✓ to opposite ends of the cell in anaphase.

This is a good answer. Mark: 3/3

Student B

(c) Mass

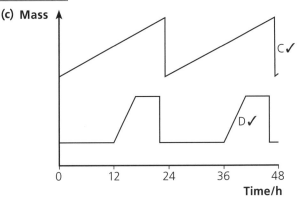

Both correct. Mark: 2/2

Student B

(d) Mitosis produces two daughter cells that are genetically identical ✓ to the parent cell. Mitosis is used for growth, or for repairing cells ✗. It is also used in asexual reproduction ✓.

The point about producing genetically identical cells is a good one, and it is also correct that mitosis is involved in asexual reproduction. However, the second sentence contains an important error. Mitosis cannot repair cells. Mitosis can produce new cells, which can help to repair tissues. Mark: 2/3

Question 4

The diagrams below show five molecules found in living organisms.

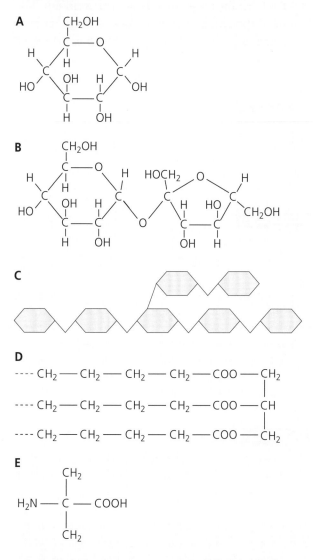

A CH₂OH

B CH₂OH

C

D
---- CH₂ — CH₂ — CH₂ — CH₂ — COO — CH₂
|
---- CH₂ — CH₂ — CH₂ — CH₂ — COO — CH
|
---- CH₂ — CH₂ — CH₂ — CH₂ — COO — CH₂

E
CH₂
|
H₂N — C — COOH
|
CH₂

(a) Give the letter of one molecule that fits each of these descriptions. You can use each letter once, more than once or not at all.
  (i) the form in which carbohydrates are transported through phloem tissue in plants **(1 mark)**
  (ii) the form in which carbohydrates are stored in animals **(1 mark)**
  (iii) a molecule that is insoluble in water **(1 mark)**
  (iv) a molecule that links together with others to form a polypeptide **(1 mark)**
  (v) a molecule that contains ester bonds **(1 mark)**

(b) Explain how the structure of water molecules makes water a good solvent. **(3 marks)**
(Total: 8 marks)

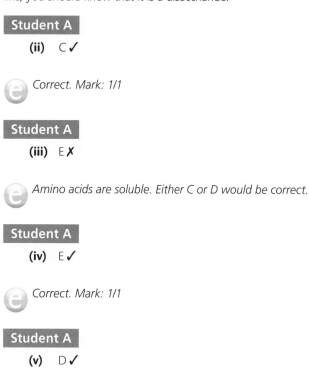

Student A

(a) (i)   A ✗

A is a glucose molecule, but plants transport sucrose. Even if you did not know what a sucrose molecule looks like, you should know that it is a disaccharide.

Student A

(ii)   C ✓

Correct. Mark: 1/1

Student A

(iii)   E ✗

Amino acids are soluble. Either C or D would be correct.

Student A

(iv)   E ✓

Correct. Mark: 1/1

Student A

(v)   D ✓

Correct. Mark: 1/1

Student A

(b) Water has dipoles and hydrogen bonds ✓, which help it to dissolve other substances.

There are no wrong statements in this answer, but it does not really give an explanation of why water is a good solvent — it just states two facts about water molecules. Mark: 1/3

Student B

(a) (i)   B ✓

Correct. Mark: 1/1

Student B

(ii)   C ✓

Correct. Mark: 1/1

Student B

(iii)   D or C ✓

Correct. However, the student took an unnecessary risk with (iii), by giving two answers. If the second one had been wrong, it could have negated the first correct one. If you are asked for one answer, it is best to give only one. *Mark: 1/1*

Student B

(iv) E ✓

Correct. *Mark: 1/1*

Student B

(v) D ✓

Correct. *Mark: 1/1*

Student B

(b) In a water molecule, the hydrogen atoms have a tiny positive electrical charge and the oxygen atom has a similar negative charge ✓. Other atoms or ions with electrical charges ✓ are attracted ✓ to these charges on the water molecules. This makes them spread about ✓ among the water molecules.

This is a good answer. It really does explain how a substance dissolves in water and relates this clearly to the structure of a water molecule. This answer contains four possible marking points, but there is a maximum of 3 marks available in total. *Mark: 3/3*

Question 5

The diagram below shows a small part of a human lung as it appears through a microscope.

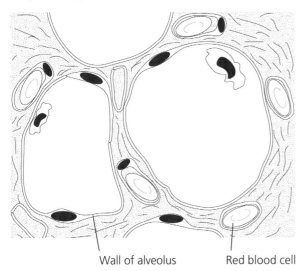

Wall of alveolus   Red blood cell

(a) Name the type of blood vessel in which the red blood cell is present.
(1 mark)

(b) Describe and explain *two* ways in which the structure of the alveoli, shown in the diagram, enables gas exchange to take place rapidly.
(4 marks)

(c) Explain why large organisms such as mammals need specialised gas exchange surfaces, whereas small organisms such as a single-celled *Amoeba* do not.
(3 marks)
(Total: 8 marks)

Student A

(a) capillary ✓

This is correct. *Mark: 1/1*

Student A

(b) They have a large surface area. ✓
They are thin, so oxygen can diffuse across quickly. ✓

The statement about a large surface area is correct, but the answer also needs to say *why* this enables gas exchange to take place rapidly (because the question asks you to 'explain'). The second answer is not sufficiently clear — what is thin? It is not the whole alveoli that are thin, but their walls. The second part of this answer does give a clear explanation of why this helps gas exchange to take place quickly. *Mark: 2/4*

Student A

(c) Large organisms have small surface areas compared with their volume ✓, so they need extra surface ✓ to be able to get enough oxygen.

There is a correct and clear statement about surface area-to-volume ratio, and the answer also just gets a second mark. However, this is not really very clear — see answer B for a better explanation. *Mark: 2/3*

Student B

(a) capillary ✓

Correct. *Mark: 1/1*

Student B

(b) Large surface area ✓ — so more oxygen and carbon dioxide molecules can diffuse across at the same time ✓.
Good supply of oxygen — to maintain a diffusion gradient between the alveoli and the blood.

The first way is correct and well explained. However, the second, although true, does not answer the question, which is about the *structure* of the alveoli. *Mark: 2/4*

Student B

(c) They have small surface area-to-volume ratios ✓, but an *Amoeba* has a large surface area-to-volume ratio. The oxygen that diffuses in across the surface has to supply the whole volume ✓ of the animal, so in a large animal that is not enough and they have specialised gas exchange surfaces to increase the surface area ✓ and let more oxygen diffuse in.

ⓔ *This is a good answer. All the important points are there and are clearly expressed. Mark: 3/3*

Question 6

The diagram below shows pressure changes in the left atrium and left ventricle of the heart and the aorta during the cardiac cycle.

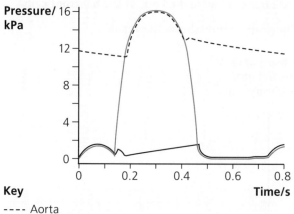

Key
---- Aorta
—— Left ventricle
—— Left atrium

(a) Calculate how many heart beats there will be in 1 minute. **(2 marks)**
(b) (i) On the diagram, indicate the point at which the semilunar valves in the aorta snap shut. **(1 mark)**
(ii) Explain what causes the semilunar valves to shut at this point in the cardiac cycle. **(2 marks)**
(iii) On the diagram, indicate the period when the left ventricle is contracting. **(1 mark)**
(iv) On the diagram, draw a line to show the changes in pressure in the right ventricle. **(2 marks)**
(c) After the blood leaves the heart, it passes into the arteries. The blood pressure gradually reduces and becomes more steady as the blood passes through the arteries.
Explain what causes this reduction and steadying of the blood pressure. **(2 marks)**
(Total: 10 marks)

Student A

(a) 1 cycle in 0.75 seconds ✓ so in 60 seconds there will be 60 × 0.75 ✗ = 45 beats.

ⓔ *The length of one cycle has been read correctly, but the calculation is wrong. Mark: 1/2*

Student A

(b) (i)

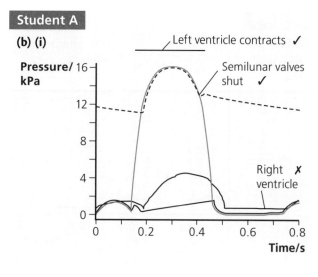

ⓔ *Correct. Mark: 1/1*

Student A

(ii) The valves shut when the ventricle starts to relax. ✓

ⓔ *This is correct as far as it goes, but more information is needed for the second mark. Mark: 1/2*

Student A

(iii) See diagram

ⓔ *Correct. Mark: 1/1*

Student A

(iv) See diagram

ⓔ *This is partly correct. The pressure in the right ventricle is correctly shown as less than that in the left ventricle, but it should be contracting and relaxing at exactly the same times as the left ventricle. Mark: 1/2*

Student A

(c) The pressure gets less as the blood gets further away from the heart ✓. The muscle in the walls of the arteries contracts ✗ and relaxes to push the blood along, and it does this in between heart beats so the pulse gets evened out.

ⓔ *The first statement is correct, but it does not really tell us any more than what is in the question. However, it is not correct that the muscles in the artery wall contract and relax to push the blood along. Mark: 1/2*

Student B

(a) $\frac{60}{0.75}$ ✓ = 80 beats per minute ✓

Correct. Mark: 2/2

Student B

(b) (i)

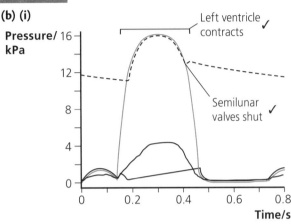

Correct. Mark: 1/1

Student B

(ii) They close when the pressure of the blood inside the arteries is higher than inside the ventricles ✓. The blood therefore pushes down on the valves and makes them shut ✓.

Correct. Mark: 2/2

Student B

(iii) See diagram

Correct. Mark: 1/1

Student B

(iv) See diagram

Correct. Mark: 2/2

Student B

(c) As the blood is forced into the artery as the ventricle contracts ✓, it pushes outwards on the artery wall, making the elastic tissue stretch ✓. In between heart beats, the pressure of the blood inside the artery falls, and the elastic tissue recoils ✓. So the wall keeps expanding and springing back. When it springs back it pushes on the blood in between ✓ heart beats, so this levels out the pressure changes.

This answer explains very well why the blood pressure levels out. However, it does not mention the overall fall in blood pressure. All the same, this is a good answer. Mark: 2/2

Question 7

Beetroot cells contain a red pigment that cannot normally escape from the cells through the cell surface membrane.

A student carried out an investigation into the effect of temperature on the permeability of the cell surface membrane of beetroot cells. She measured permeability by using a colorimeter to measure the absorbance of green light by the solutions in which samples of beetroot had been immersed. The greater the absorbance, the more red pigment had leaked out of the beetroot cells

The graph below shows her results.

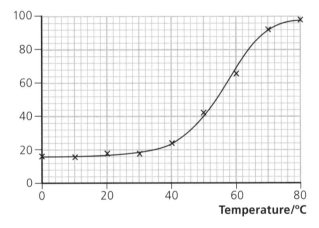

(a) With reference to the graph, describe the effect of temperature on the absorbance of light in the colorimeter. (3 marks)

(b) With reference to the structure of cell membranes, explain the effects you have described in (a). (4 marks)

(Total: 7 marks)

Student A

(a) Between 0 and 30 the absorbance goes up very slightly ✓. Above 40 °C it goes up very quickly ✗. Then it starts to level out at about 70 °C ✓.

The three main regions of the graph have been identified correctly, stating where changes in gradient occur. However, the term 'quickly' is used, which is not correct because the graph does not show anything about time. A third mark could have been gained by quoting some figures from the graph. Mark: 2/3

Student A

(b) High temperatures damage the proteins in the membrane ✓. They become denatured, so they leave holes ✓ in the membrane that the beetroot pigment can get through.

 This is a good answer as far as it goes, and is clearly expressed. However, it is not enough to score full marks. Mark: 2/4

Student B

(a) The general trend is that the higher the temperature, the greater the absorbance. Between 0 and 30°C, the absorbance increases very slightly ✓ from about 16 to 18 arbitrary units. Above 40°C it increases much more steeply ✓, levelling off at about 70°C ✓. The maximum absorbance is 98 arbitrary units.

 This is a good answer. However, although some figures from the graph are quoted, the student has not manipulated them in any way — for example, the increase in absorbance between 0 and 30°C could have been calculated. Mark: 3/3

Student B

(b) As temperature increases, the phospholipids and protein molecules in the membrane move about faster ✓ and with more energy. This leaves gaps in the membrane, so the beetroot pigment molecules can get through ✓ and escape from the cell. The protein molecules start to lose their shape at high temperatures ✓ because their hydrogen bonds break ✓, so the protein pores get wider ✓, which increases permeability.

 This is an excellent answer. Mark: 4/4

12 Energy and respiration

Energy in living organisms

Revised

ATP

ATP stands for **adenosine triphosphate**. ATP is a phosphorylated nucleotide — it has a similar structure to the nucleotides that make up RNA. However, it has three phosphate groups attached to it instead of one (Figure 12.1).

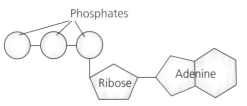

Figure 12.1 An ATP molecule

ATP is used as the energy currency in every living cell. When an ATP molecule is hydrolysed, it loses one of its phosphate groups and some energy is released, which can be used by the cell. In this process, the ATP is converted to ADP (adenosine diphosphate — Figure 12.2).

Figure 12.2 Hydrolysis of ATP and formation of ADP

Cells use energy for many different purposes. These include:

● the synthesis of proteins and other large molecules from smaller ones, including DNA replication. These are examples of **anabolic reactions** — that is, energy-consuming reactions.
● the active transport of ions and molecules across cell membranes against their concentration gradient (p. 37)
● the transmission of nerve impulses (p. 130)
● movement, for example muscle contraction (such as heartbeat, breathing movements, walking) or movement of cilia
● the production of heat to maintain body temperature at a steady level (in mammals and birds)

Each cell makes its own ATP. The hydrolysis of one ATP molecule releases a small 'packet' of energy that is often just the right size to fuel a particular step in a process. A glucose molecule, on the other hand, would contain far too much energy, so a lot would be wasted if cells used glucose molecules as their immediate source of energy.

All cells make ATP by respiration. This is described in the next few pages.

Respiration

All cells obtain useable energy through respiration. Respiration is the oxidation of energy-containing organic molecules, such as glucose. These are known as **respiratory substrates**. The energy released from this process is used to combine ADP with inorganic phosphate (P_i) to make ATP.

Respiration can take place in **aerobic** or **anaerobic conditions**. In both cases, glucose or another respiratory substrate is oxidised.

In aerobic respiration, oxygen is involved, and the substrate is oxidised completely, releasing much of the energy that it contains.

In anaerobic conditions, respiration takes place without oxygen, and the substrate is only partially oxidised. Only a small proportion of the energy it contains is released.

> **Respiration** is a series of enzyme-controlled reactions that takes place in cells, in which energy within a nutrient molecule (e.g. glucose) is used to make ATP.

> **Typical mistake**
>
> Do not say that respiration 'produces' or 'makes' energy. The energy is already there, in the glucose molecule.

Coenzymes
Revised

Respiration involves **coenzymes** called NAD, FAD and coenzyme A. A coenzyme is a molecule required for an enzyme to be able to catalyse a reaction. NAD and FAD are reduced during respiration. The term 'reduce' means to add hydrogen, so reduced NAD has had hydrogen added to it. Without the presence of NAD or FAD to accept the hydrogen, the dehydrogenase enzymes involved in respiration would not be able to remove hydrogen from their substrates.

Aerobic respiration
Revised

Glucose, $C_6H_{12}O_6$ (or another respiratory substrate) is split to release carbon dioxide as a waste product. The hydrogen from the glucose is combined with atmospheric oxygen. This releases a large amount of energy, which is used to drive the synthesis of ATP.

Glycolysis
Glycolysis is the first stage of respiration. It takes place in the cytoplasm.

- A glucose (or other hexose sugar) molecule is phosphorylated, as two ATPs donate phosphate to it.
- This produces a **fructose 1,6-bisphosphate** molecule (6C), which splits into two **triose phosphate** molecules.
- Each triose phosphate is oxidised to form a **pyruvate** molecule. This involves the removal of hydrogens by dehydrogenase enzymes, which are taken up by the **coenzyme NAD**. The removal of hydrogens is an **oxidation reaction**. It can also be referred to as **dehydrogenation**. This produces **reduced NAD**. During this step, the phosphate groups from the triose phosphates are added to ADP to produce a small yield of ATP, in substrate-linked reactions.
- Overall, two molecules of ATP are used and four are made during glycolysis of one glucose molecule, making a net gain of two ATPs per glucose (Figure 12.3).

The link reaction
If oxygen is available, each pyruvate now moves into a mitochondrion (Figure 12.4), where the **link reaction** and the **Krebs cycle** take place. During these processes, the glucose is completely oxidised.

> **Now test yourself**
>
> 1 Suggest why glucose must be phosphorylated before the 6C molecule can be split into two 3C molecules.
>
> **Answer on p.203**
>
> Tested

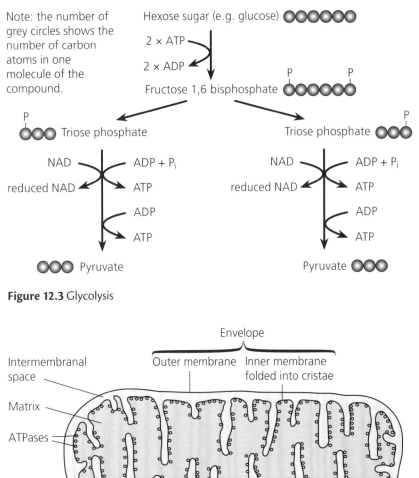

Note: the number of grey circles shows the number of carbon atoms in one molecule of the compound.

Hexose sugar (e.g. glucose)

2 × ATP

2 × ADP

Fructose 1,6 bisphosphate

Triose phosphate

NAD — reduced NAD — ADP + P_i — ATP — ADP — ATP

Triose phosphate

NAD — reduced NAD — ADP + P_i — ATP — ADP — ATP

Pyruvate

Pyruvate

Figure 12.3 Glycolysis

Envelope

Intermembranal space

Outer membrane   Inner membrane folded into cristae

Matrix

ATPases

0.1 μm

Figure 12.4 A mitochondrion

Carbon dioxide is removed from the pyruvate. This is called **decarboxylation**. This carbon dioxide diffuses out of the mitochondrion and out of the cell.

Hydrogen is also removed from the pyruvate, and is picked up by NAD, producing reduced NAD. This converts pyruvate into a 2C compound. This immediately combines with coenzyme A to produce **acetyl coenzyme A,** which is also a 2C compound.

The Krebs cycle

We have seen that acetyl coenzyme A has two carbon atoms. It combines with a **4C** compound called **oxaloacetate** to produce a **6C** compound, **citrate**. The citrate is gradually converted to the 4C compound again through a series of enzyme-controlled steps (Figure 12.5). These steps all take place in the matrix of the mitochondrion, and each is controlled by specific enzymes.

During this process:

● more carbon dioxide is released (decarboxylation) and diffuses out of the mitochondrion and out of the cell

● more hydrogens are released (dehydrogenation) and picked up by NAD and another coenzyme called FAD; this produces reduced NAD and reduced FAD

● some ATP is produced from ADP, in substrate-linked reactions

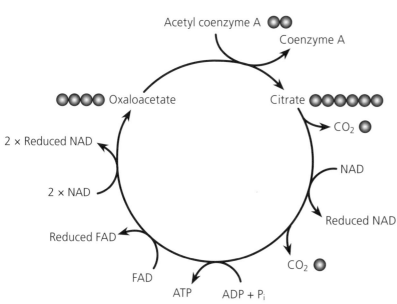

Figure 12.5 The Krebs cycle

Oxidative phosphorylation

The hydrogens picked up by NAD and FAD are now split into electrons and protons. The electrons are passed along the **electron transport chain**, on the inner membrane of the mitochondrion.

As the electrons move along the chain, they release energy. This energy is used to actively transport protons (hydrogen ions) from the matrix of the mitochondrion, across the inner membrane and into the space between the inner and outer membranes. This builds up a high concentration of protons in this space.

The protons are allowed to move back by facilitated diffusion into the matrix through special channel proteins that work as ATP synthases. The movement of the protons through the ATP synthase provides enough energy to cause ADP and inorganic phosphate to combine to make ATP.

At the end of the chain, the electrons reunite with the protons from which they were originally split. They combine with oxygen to produce water (Figure 12.6). This is why oxygen is required in aerobic respiration — it acts as the final acceptor for the hydrogens removed from the respiratory substrate during glycolysis, the link reaction and the Krebs cycle.

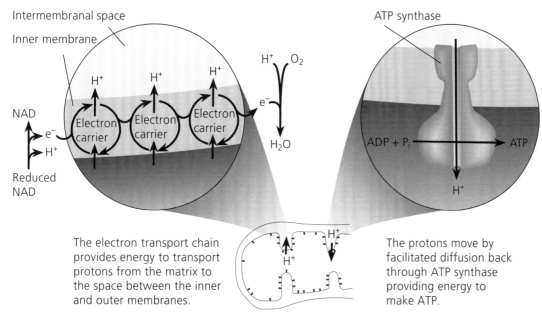

Figure 12.6 Oxidative phosphorylation

> ### Now test yourself
>
> **2** How many protons and electrons are there in one hydrogen atom?
>
> **Answer on p.203**
>
> Tested

The relationship between structure and function of mitochondria

Look back at Figure 12.4 on p. 102.

- A mitochondrion is surrounded by an envelope (two membranes) that separate it from the cytoplasm, so that the reactions that take place inside it are not affected by reactions elsewhere in the cell.
- The inner membrane is folded to form cristae, providing a large area in which the carriers of the electron transport chain, and ATP synthase, can be embedded.
- The space between the two membranes (intermembrane space) is available to build up a high concentration of protons.
- The matrix of the mitochondrion contains all the enzymes needed for the Krebs cycle.
- The matrix of the mitochondrion also contains DNA and ribosomes, which are used to synthesise some of the proteins needed for the reactions of respiration.

Respiration in anaerobic conditions

If oxygen is not available, oxidative phosphorylation cannot take place, because there is nothing to accept the electrons and protons at the end of the electron transport chain. This means that reduced NAD is not reoxidised, so the mitochondrion quickly runs out of NAD or FAD that can accept hydrogens from the Krebs cycle reactions. The Krebs cycle and the link reaction therefore come to a halt.

Glycolysis, however, can still continue, so long as the pyruvate produced at the end of it can be removed and the reduced NAD can be converted back to NAD.

The lactate pathway

In mammals, the pyruvate is removed by converting it to **lactate** (Figure 12.7).

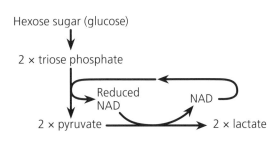

Figure 12.7 The lactate pathway

The lactate that is produced (usually in muscles) diffuses into the blood and is carried in solution in the blood plasma to the liver. Here, liver cells convert it back to pyruvate. This requires oxygen, so extra oxygen is required after exercise has finished. The extra oxygen is known as the **oxygen debt**. Later, when the exercise has finished and oxygen is available again, some of the pyruvate in the liver cells is oxidised through the link reaction, the Krebs cycle and the electron transport chain. Some of the pyruvate is reconverted to glucose in the liver cells. The glucose may be released into the blood or converted to glycogen and stored.

The ethanol pathway

In yeast and in plants, the pyruvate is removed by converting it to ethanol (Figure 12.8, overleaf).

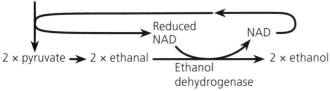

Figure 12.8 The ethanol pathway

Adaptations of rice for wet conditions

Rice is often grown in fields that are flooded with water during the growing season. The roots therefore have little oxygen available to them. Rice has adaptations that allow it to grow successfully even when respiration in root cells takes place in these anaerobic conditions (Table 12.1).

Table 12.1 Adaptations of rice for growth with roots submerged in water

Feature	How it helps the plant to survive when roots are submerged
Cells are tolerant of high concentrations of ethanol	When roots are submerged in water, less oxygen is available than when the soil contains air spaces; cells therefore respire anaerobically, producing ethanol
Stems have tissue called aerenchyma, containing large air spaces	Aerenchyma allows oxygen from the air to diffuse down to the roots
Some types of rice are able to grow elongated stems to keep their leaves above the water as its level rises	The leaves remain exposed to the air, which facilitates gas exchange for photosynthesis and respiration

ATP yield in aerobic and anaerobic respiration

Revised

Only small amounts of ATP are produced when one glucose molecule undergoes anaerobic respiration. This is because only glycolysis is completed. The Krebs cycle and oxidative phosphorylation, which produce most ATP, do not take place.

The precise number of molecules of ATP produced in aerobic respiration of one glucose molecule varies between different organisms and different cells, but is usually between 30 and 32 molecules. Figure 12.9 summarises the four stages in aerobic respiration, and also shows ATP yields of each stage.

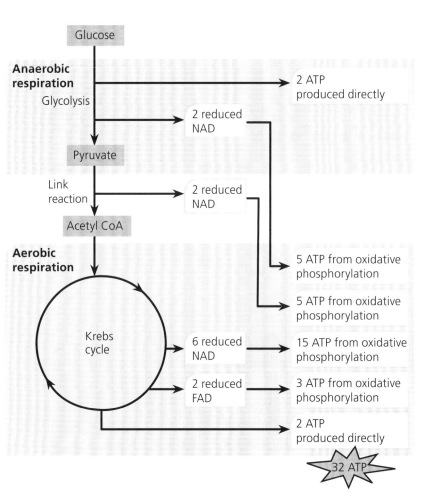

Figure 12.9 A summary of the four stages of respiration

Respiratory substrates

Glucose is not the only respiratory substrate. All carbohydrates, lipids and proteins can also be used as respiratory substrates (Table 12.2). Lipid provides more than twice as much energy per gram as carbohydrate or protein. This is because a lipid molecule contains relatively more hydrogen atoms (in comparison with carbon or oxygen atoms) than carbohydrate or protein molecules do. You have seen that it is hydrogen atoms that provide the protons that are used to generate ATP via the electron transport chain.

Table 12.2 Energy values of different respiratory substrates

Respiratory substrate	Energy released/kJg^{-1}
Carbohydrate	16
Lipid	39
Protein	17

Many cells in the human body are able to use a range of different respiratory substrates. However, brain cells can only use glucose. Heart muscle preferentially uses fatty acids.

Respiratory quotients

It is possible to get a good idea of which respiratory substrate the cells in an organism are using by measuring the volume of oxygen it is taking in and the volume of carbon dioxide it is giving out.

The **respiratory quotient**, **RQ** is $\dfrac{\text{volume of } CO_2 \text{ given out}}{\text{volume of } O_2 \text{ taken in}}$

The values in Table 12.3 are for aerobic respiration. If a cell or an organism is respiring in anaerobic conditions, then no oxygen is being used. The RQ is therefore infinity (∞).

Typical mistake

Do not confuse RQ with R_f values (see p. 112).

Table 12.3 RQs for different substrates undergoing aerobic respiration

Respiratory substrate	RQ
Carbohydrate	1.0
Lipid	0.7
Protein	0.9

How to... **Use respirometers**

There are different types of respirometer. One type is shown in Figure 12.10.

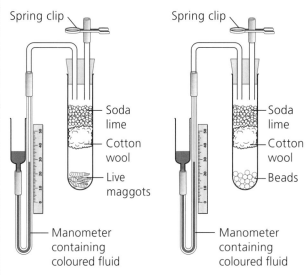

Figure 12.10 A respirometer

Using a respirometer to measure the rate of uptake of oxygen

The organisms to be investigated are placed in one tube, and non-living material of the same mass in the other tube. Soda lime is placed in each tube, to absorb all carbon dioxide. Cotton wool prevents contact of the soda lime with the organisms.

Coloured fluid is poured into the reservoir of each manometer and allowed to flow into the capillary tube. It is essential that there are no air bubbles. You must end up with exactly the same quantity of fluid in the two manometers.

Two rubber bungs are now fitted with tubes as shown in the diagram. Close the spring clips. Attach the manometers to the bent glass tubing, ensuring an airtight connection. Next, place the bungs into the tops of the tubes.

Open the spring clips. (This allows the pressure throughout the apparatus to equilibrate with atmospheric pressure.) Note the level of the manometer fluid in each tube. Close the clips. Each minute, record the level of the fluid in each tube.

As the organisms respire, they take oxygen from the air around them and give out carbon dioxide. The removal of oxygen from the air inside the tube reduces the volume and pressure, causing the manometer fluid to move towards the organisms.

The carbon dioxide given out is absorbed by the soda lime. The distance moved by the fluid is therefore affected only by the oxygen taken up and not by the carbon dioxide given out.

You would not expect the manometer fluid in the tube with no organisms to move, but it may do so because of temperature changes. This allows you to control for this variable, by subtracting the distance moved by the fluid in the control manometer from the distance moved in the experimental manometer (connected to the living organisms), to give you an adjusted distance moved.

Calculate the mean (adjusted) distance moved by the manometer fluid per minute. If you know the diameter of the capillary tube, you can convert the distance moved to a volume:

volume of liquid in a tube = length × πr^2

This gives you a value for the volume of oxygen absorbed by the organisms per minute.

Using a respirometer to investigate the effect of temperature on the rate of respiration

The respirometer can be placed in water baths at different temperatures. You can use the same respirometer for the whole experiment. Or you could have different ones for each temperature. (In each case, there are difficulties with controlling some variables — you might like to think about what these are.) At each temperature, you need a control respirometer with no organisms in it.

If you are simply *comparing* the rates of respiration at different temperatures, then you do not need to convert the distance moved by the manometer fluid to a volume. You could just plot distance moved on the *y*-axis of your graph and time on the *x*-axis.

The rate of respiration is represented by the gradient of the graph (Figure 12.11).

At 50 °C manometer fluid travelled 47 mm in 50 s

Rate of respiration = $\dfrac{47}{50}$ = 0.94 mm s^{-1}

At 30 °C manometer fluid travelled 40 mm in 60 s

Rate of respiration = $\dfrac{40}{60}$ = 0.67 mm s^{-1}

At 20 °C manometer fluid travelled 21 mm in 70 s

Rate of respiration = $\dfrac{24}{70}$ = 0.34 mm s^{-1}

→

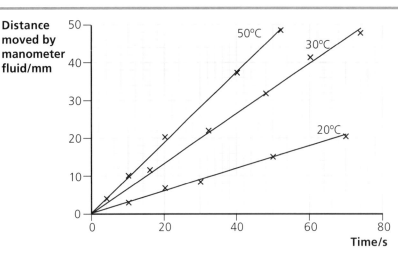

Figure 12.11 Comparing rates of respiration at different temperatures

Using a respirometer to measure RQ

For this, we need to know both how much oxygen is taken in, and how much carbon dioxide is given out.

Set up two respirometers as shown in Figure 12.10. However, the second respirometer should also contain the same mass of live maggots (or whatever organism you are investigating) but should *not* contain soda lime. You could put some inert material into the tube (for example, the beads) instead of soda lime. The mass and volume of the inert material should be the same as the mass and volume of the soda lime.

This second tube is therefore just like the first one except that it does not contain soda lime. The carbon dioxide given out by the respiring organisms is therefore not absorbed.

The difference between the distance moved by the manometer fluid in the experimental tube and the distance moved in the control tube is therefore due to the carbon dioxide given out.

distance moved by fluid in experimental tube = x mm

distance moved by fluid in control tube = y mm

x mm represents the oxygen taken up

$x - y$ represents the carbon dioxide given out

therefore RQ = $\dfrac{x - y}{x}$

For example, if the respiratory substrate is carbohydrate, then the amount of carbon dioxide given out will equal the amount of oxygen taken in. The fluid in the control tube will not move, so $y = 0$.

RQ is then $\dfrac{x - 0}{x} = 1$

Now test yourself
Tested

3 Predict and explain what would happen to the levels of fluid in the manometers if no soda lime was used.

Answer on p.203

How to... Use a redox indicator to investigate rate of respiration in yeast

- We have seen that respiration involves stages in which hydrogen is removed from substrate, a process called dehydrogenation. We can investigate the rate of dehydrogenase reactions using an indicator that can accept the hydrogens removed in respiration, and that changes colour when it is reduced (has hydrogens added) or oxidised (has hydrogens removed). Suitable indicators include methylene blue and DCPIP. Both of these indicators are blue when oxidised, and become colourless when they are reduced.

- You can place a suspension of yeast in glucose solution into two boiling tubes. To one of the tubes, add your indicator. Note the time taken for the indicator to change colour. As the yeast suspension is not colourless, it is helpful to compare the colour of the tube with the indicator with the tube to which you did not add indicator.

- You can use this technique to investigate the effect of temperature or substrate concentration on the rate of respiration. To vary temperature, use water baths. To vary substrate concentration, change the amount of glucose in the yeast suspension. Measure the time taken for the indicator to become colourless.

13 Photosynthesis

An overview of photosynthesis

Photosynthesis is a series of reactions in which energy transferred as light is transformed to chemical energy. Energy from light is trapped by chlorophyll, and this energy is then used to:

- split apart the strong bonds in water molecules to release hydrogen
- produce ATP
- reduce a substance called NADP

NADP stands for nicotinamide adenine dinucleotide phosphate, which — like NAD — is a coenzyme.

The ATP and reduced NADP are then used to add hydrogen to carbon dioxide, to produce carbohydrate molecules such as glucose (Figure 13.1). These complex organic molecules contain some of the energy that was originally in the light. The oxygen from the split water molecules is a waste product, and is released into the air.

There are many different steps in photosynthesis, which can be divided into two main stages — the light dependent stage and the light independent stage.

Typical mistake

Students often say that carbon dioxide is changed to oxygen in photosynthesis. Look at Figure 13.1 and see for yourself why this is not correct.

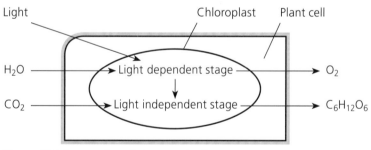

Figure 13.1 An overview of photosynthesis

Chloroplasts
Revised

Photosynthesis takes place inside **chloroplasts**. Like mitochondria, these are organelles surrounded by two membranes, called an **envelope** (Figure 13.2). They are found in mesophyll cells in leaves. Palisade mesophyll cells contain most chloroplasts but they are also found in spongy mesophyll cells. Guard cells also contain chloroplasts. You can see a diagram of the structure of a leaf on p. 117.

The membranes inside a chloroplast are called **lamellae**, and it is here that the light dependent stages take place. The membranes contain **chlorophyll** molecules, arranged in groups called **photosystems**. There are two kinds of photosystem, PSI and PSII, each of which contains slightly different kinds of chlorophyll.

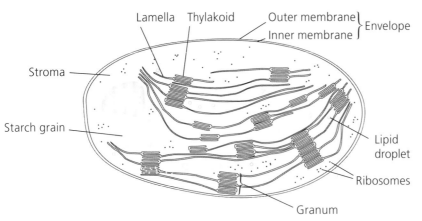

Figure 13.2 Structure of a chloroplast

There are enclosed spaces between pairs of membranes, forming fluid-filled sacs called **thylakoids**. These are involved in photophosphorylation — the formation of ATP using energy from light. Thylakoids are often arranged in stacks called **grana** (singular: granum).

The 'background material' of the chloroplast is called the **stroma**, and this is where the light independent stage takes place.

Chloroplasts often contain starch grains and lipid droplets. These are stores of energy-containing substances that have been made in the chloroplast but are not immediately needed by the cell or by other parts of the plant.

Revision activity

● Compare the structures of a mitochondrion and a chloroplast.

Chloroplast pigments

Revised

A **pigment** is a substance that absorbs light of some wavelengths but not others. The wavelengths that is does **not** absorb are reflected from it.

Chlorophyll is the main pigment contained in chloroplasts. It looks green because it reflects green light. Other wavelengths (colours) of light are absorbed.

Figure 13.3 shows the wavelengths of light absorbed by the various pigments found in chloroplasts. These graphs are called **absorption spectra**.

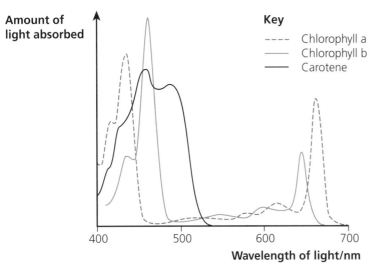

Figure 13.3 Absorption spectra for chloroplast pigments

If we shine light of various wavelengths on chloroplasts, we can measure the rate at which they give off oxygen. This graph is called an **action spectrum** (see Figure 13.4).

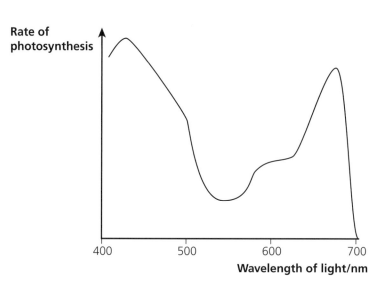

Figure 13.4 Action spectrum for chloroplast pigments

Now test yourself

1 Explain the similarities between the absorption spectrum and the action spectrum.

Answer on p.203

Tested

Chlorophyll a is the most abundant pigment in most plants. Its absorption peaks are 430 nm (blue) and 662 nm (red). It emits an electron when it absorbs light.

Chlorophyll b is similar to chlorophyll a, but its absorption peaks are 453 nm and 642 nm. It has a similar role to chlorophyll a, but is not as abundant.

Carotenoids are **accessory pigments**. They are orange pigments that protect chlorophyll from damage by the formation of single oxygen atoms (free radicals). They can also absorb wavelengths of light that chlorophyll cannot absorb, and pass on some of the energy from the light to chlorophyll.

Xanthophylls are also accessory pigments, capturing energy from wavelengths of light that are not absorbed by chlorophyll.

How to... Separate chlorophyll pigments by chromatography

Chromatography is a method of separation that relies on the different solubilities of different solutes in a solvent. A mixture of chlorophyll pigments is dissolved in a solvent, and then a small spot is placed onto chromatography paper. The solvent gradually moves up the paper, carrying the solutes with it. The more soluble the solvent, the further up the paper it is carried.

There are various methods. The one described in Figure 13.5 uses thin layer chromatography on specially prepared strips instead of paper. Only an outline of the procedure is given here, so you cannot use these instructions to actually carry out the experiment. You can find more details about this technique at **www.saps.org.uk/secondary/teaching-resources/181-student-sheet-10-thin-layer-chromatography-for-photosynthetic-pigments**. Your teacher should give you an opportunity to do a chromatography experiment, perhaps using a different technique from the one described here.

Cut a TLC plate into narrow strips, about 1.25 cm wide, so they fit into a test tube. Do not put your fingers on the powdery surface.

Put 2 or 3 grass leaves on a slide. Using another slide scrape the leaves to extract cell contents.

Add 6 drops of propanone to the extract and mix.

Figure 13.5

Transfer the mixture to a watch glass. Allow this to dry out almost completely — a warm air flow will speed this up.

Transfer tiny amounts of the concentrated extract onto a spot 1 cm from one end of the TLC strip.

Touch very briefly with the fine tip of the brush and let that spot dry before adding more. Keep the spot to 1 mm diameter if you can. The final spot, called the origin, should be small but dark green.

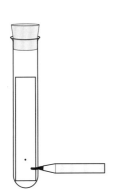 

Put the TLC strip in a test tube. Mark the tube below the pigment spot and remove the TLC strip.

Add the running solvent to the depth of the mark, then return the TLC strip to the tube and seal it.

After about 4 minutes remove the TLC strip. Immediately mark the solvent front with a needle. You can also mark the centres of the pigment spots and the origin.

- Measure the distance from the start line to the solvent front. Measure the distances of each pigment spot from the start line. For each spot, calculate the R_f value:

$$R_f = \frac{\text{distance from start line to pigment spot}}{\text{distance from start line to solvent front}}$$

- You can use the R_f values to help you to identify the pigments. R_f values differ depending on the solvent you have used, but typical values might be:

chlorophyll a	0.60
chlorophyll b	0.50
carotene	0.95
xanthophyll	0.35

You may also see a small grey spot with an R_f value of about 0.8. This is phaeophytin, which is not really a chlorophyll pigment, but is a breakdown product generated during the extraction process.

The light dependent stage
Revised

Chlorophyll molecules in PSI and PSII absorb light energy. The energy excites electrons, raising their energy level so that they leave the chlorophyll. The chlorophyll is said to be photoactivated.

PSII contains an enzyme that splits water when activated by light. This reaction is called **photolysis** ('splitting by light'). The water molecules are split into oxygen and hydrogen atoms. Each hydrogen atom then loses its electron, to become a positively charged hydrogen ion (proton), H^+. The electrons are picked up by the chlorophyll in PSII, to replace the electrons they lost. The oxygen atoms join together to form oxygen molecules, which diffuse out of the chloroplast and into the air around the leaf.

$$2H_2O \xrightarrow{\text{light}} 4H^+ + 4e^- + O_2$$

The electrons emitted from PSII are picked up by **electron carriers** (electron transport chain) in the membranes of the thylakoids. They are passed along a

> The **light dependent stage** of photosynthesis is the series of events that only occurs when light is present; it involves the use of energy from light to split water molecules and produce ATP and reduced NADP.

chain of these carriers, losing energy as they go. The energy they lose is used to make ADP combine with a phosphate group, producing ATP. This is called **photophosphorylation**. At the end of the electron carrier chain, the electron is picked up by PSI, to replace the electron the chlorophyll in PSI had lost.

The electrons from PSI are passed along a different chain of carriers to NADP. The NADP also picks up the hydrogen ions from the split water molecules. The NADP becomes reduced NADP.

We can show all of this in a diagram called the **Z-scheme** (Figure 13.6). The higher up the diagram, the higher the energy level. If you follow one electron from a water molecule, you can see how it:

- is taken up by PSII
- has its energy raised as the chlorophyll in PSII absorbs light energy
- loses some of this energy as it passes along the electron carrier chain
- is taken up by PSI
- has its energy raised again as the chlorophyll in PSI absorbs light energy
- becomes part of a reduced NADP molecule

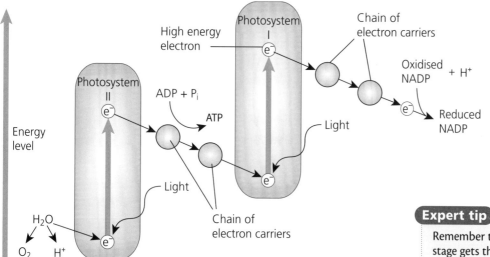

Figure 13.6 Summary of the light dependent stage of photosynthesis — the Z-scheme

At the end of this process, two new substances have been made. These are ATP and reduced NADP. Both of them will now be used in the next stage of photosynthesis, the light independent stage.

Non-cyclic and cyclic photophosphorylation

The sequence of events just described and shown in Figure 13.6 is known as **non-cyclic photophosphorylation**.

There is an alternative pathway for the electron that is emitted from PSI. It can simply be passed along the electron transport chain, then back to PSI again. ATP is produced as it moves along the electron transport chain (photophosphorylation). However, no reduced NADP is produced. This is called **cyclic photophosphorylation**.

The light independent stage
Revised

The light independent stage is made up a cycle of reactions known as the **Calvin cycle** (Figure 13.7). It takes place in the stroma of the chloroplast, where the enzyme ribulose bisphosphate carboxylase, usually known as **rubisco**, is found.

Expert tip

Remember that the light dependent stage gets the energy to drive the reactions from light. It is therefore not affected by temperature.

Now test yourself

2 The Z-scheme shows that electrons lose energy as they pass along the chain of electron carriers. Where does this energy go?

Answer on p.203

Tested

Typical mistake

Do not call the light independent stage 'the dark stage'. Although this stage does not need light, it can take place in the light.

Carbon dioxide diffuses into the stroma from the air spaces within the leaf. It enters the active site of rubisco, which combines it with a 5C compound called ribulose bisphosphate, **RuBP**. The products of this reaction are two 3C molecules, glycerate 3-phosphate, **GP**. The combination of carbon dioxide with RuBP is called **carbon fixation**.

Energy from ATP and hydrogen from reduced NADP are then used to reduce the GP to **triose phosphate**, **TP**. Triose phosphate is the first carbohydrate produced in photosynthesis.

Most of the triose phosphate is used to regenerate ribulose bisphosphate, so that more carbon dioxide can be fixed. The rest is used to make glucose or whatever other organic substances the plant cell requires. These include polysaccharides such as starch for energy storage and cellulose for making cell walls, sucrose for transport, amino acids for making proteins, lipids for energy storage and nucleotides for making DNA and RNA.

> The **light independent stage** of photosynthesis is a series of reactions that can take place even when light is not present; it uses ATP and NADP from the light dependent stage to synthesise carbohydrates from carbon dioxide.

Now test yourself

3 What are the substrate and product of rubisco?

4 Where do the ATP and NADP used in the Calvin cycle come from?

Answers on p.203

Tested

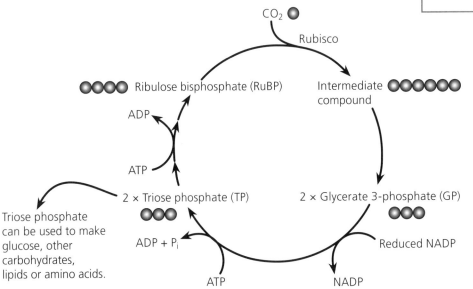

Figure 13.7 The Calvin cycle

Limiting factors in photosynthesis

The rate at which photosynthesis takes place is directly affected by several environmental factors.

- **Light intensity**. This affects the rate of the light dependent stage, because this is driven by energy transferred in light rays.
- **Temperature**. This affects the rate of the light independent stage. At higher temperatures, molecules have more kinetic energy so collide more often and are more likely to react when they do collide. (The rate of the light dependent stage is not affected by temperature, as the energy that drives this process is light energy, not heat energy.)
- **Carbon dioxide concentration** in the atmosphere. Carbon dioxide is a reactant in photosynthesis. Normal air contains only about 0.04% carbon dioxide.
- **Availability of water**. Water is a reactant in photosynthesis, but there is usually far more water available than carbon dioxide, so even if water supplies are low this is not usually a problem. However, water supply can affect the rate of photosynthesis *indirectly*, because a plant that is short of water will close its stomata, preventing carbon dioxide from diffusing into the leaf.

If the level of any one of these factors is too low, then the rate of photosynthesis will be reduced. The factor that has the greatest effect in reducing the rate is said to be the **limiting factor** (Figure 13.8).

A **limiting factor** is a factor that, when in short supply, limits the rate of a reaction or process.

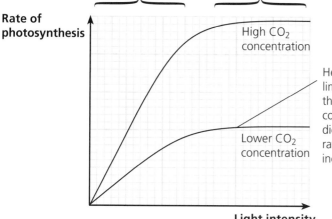

Over this range of the graphs, light intensity is the limiting factor. If light intensity increases, then the rate of photosynthesis increases.

Over this range of the graph, light intensity is not a limiting factor. If light intensity increases, there is no effect on the rate of photosynthesis. Some other factor is limiting the rate.

Rate of photosynthesis

High CO_2 concentration

Lower CO_2 concentration

Here, carbon dioxide is the limiting factor. We can tell this is so because when the concentration of carbon dioxide is increased, the rate of photosynthesis increases (see top curve).

Light intensity

Figure 13.8 Limiting factors for photosynthesis

Commercial plant growers can manipulate these factors to maximise the rate of photosynthesis and therefore increase the yields of their crops. They can:
- grow crops in glasshouses, where they can control light intensity, temperature and carbon dioxide concentration
- cover crops growing in open fields with transparent plastic (e.g. as polytunnels) to increase temperature
- irrigate crops growing in open fields

How to... Investigate the effect of environmental factors on the rate of photosynthesis of whole plants

One way to measure the rate of photosynthesis is to measure the rate at which oxygen is given off by an aquatic plant such as *Elodea* or *Cabomba*. There are various ways in which oxygen can be collected and measured. One method is shown in Figure 13.9.

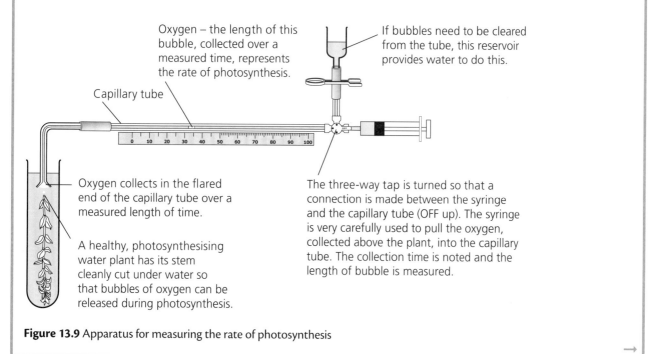

Oxygen – the length of this bubble, collected over a measured time, represents the rate of photosynthesis.

If bubbles need to be cleared from the tube, this reservoir provides water to do this.

Capillary tube

Oxygen collects in the flared end of the capillary tube over a measured length of time.

A healthy, photosynthesising water plant has its stem cleanly cut under water so that bubbles of oxygen can be released during photosynthesis.

The three-way tap is turned so that a connection is made between the syringe and the capillary tube (OFF up). The syringe is very carefully used to pull the oxygen, collected above the plant, into the capillary tube. The collection time is noted and the length of bubble is measured.

Figure 13.9 Apparatus for measuring the rate of photosynthesis

Alternatively, you can make calcium alginate balls containing green algae and place them in hydrogencarbonate indicator solution. As the algae photosynthesise, they take in carbon dioxide, which causes the pH around them to increase. The indicator changes from orange, through red to magenta.

You can find details of this technique at **www.saps.org. uk/secondary/teaching-resources/235-student-sheet-23-photosynthesis-using-algae-wrapped-in-jelly-balls**.

Whichever technique is used, you should change one factor (your independent variable) while keeping all others constant (the control variables). The dependent variable will be the rate at which oxygen is given off (measured by the volume of oxygen collected per minute in the capillary tube) *or* the rate at which carbon dioxide is used (measured by the rate of change of colour of the hydrogencarbonate indicator solution).

You could investigate the following independent variables:
- Light intensity. You can vary this by using a lamp to shine light onto the plant or algae. The closer the lamp, the higher the light intensity.
- Wavelength of light. You can vary this by placing coloured filters between the light source and the plant. Each filter will allow only light of certain wavelengths to pass through.
- Carbon dioxide concentration. You can vary this by adding sodium hydrogencarbonate to the water around the aquatic plant. This contains hydrogencarbonate ions, which are used as a source of carbon dioxide by aquatic plants.
- Temperature. The part of the apparatus containing the plant or algae can be placed in a water bath at a range of controlled temperatures.

Expert tip

It is not easy to vary light intensity without also varying temperature, because the light from a lamp also heats the water. You can try putting a piece of transparent plastic between the light and the water.

How to...   Investigate the rate of photosynthesis using a redox indicator

Photosynthesis, like respiration, involves the acceptance of hydrogen by a coenzyme. This occurs during the light-dependent stage, when hydrogen is accepted by NADP. We can investigate the rate at which this occurs by adding a redox indicator, such as DCPIP, to a suspension of chloroplasts. The indicator takes up the hydrogen ions that are produced as the light dependent stage occurs in the chloroplasts, and loses its colour. This is called the Hill reaction. The rate at which the

colour is lost is determined by the rate of the light dependent stage.

You can find a detailed protocol for carrying out this investigation at **www.nuffieldfoundation.org/practical-biology/investigating-light-dependent-reaction-photosynthesis**.

You can use this technique to investigate the effect of light intensity or light wavelength on the rate of photosynthesis. See above for methods of altering these two independent variables.

A **redox indicator** is a substance that changes colour when it is oxidised or reduced.

Leaf structure in C3 and C4 plants

Plants in which photosynthesis takes place as described are called **C3 plants**. This is because the first compound that is produced in the Calvin cycle (GP) contains 3 carbon atoms. Figure 13.10 shows how various features of a leaf of a C3 plant are adaptations that enable photosynthesis to take place.

Some plants, such as the crops maize and sorghum, are **C4 plants**. This means that, instead of first making a 3C compound during the Calvin cycle, they produce a 4C compound. This is an adaptation to growing in environments where the temperature and light intensity are high.

At high temperatures and high light intensities, the enzyme rubisco tends to catalyse the combination of RuBP with oxygen rather than with carbon dioxide. This is wasteful, and reduces the rate of photosynthesis. It is sometimes called 'photorespiration', because it uses oxygen. The leaves of C4 plants have structural adaptations that prevent photorespiration from taking place.

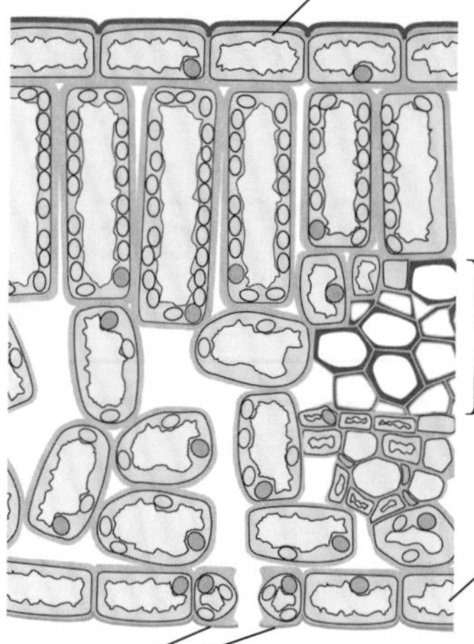

The overall shape of most leaves is thin and broad. The thinness allows sunlight to pass through to all the palisade and spongy cells. The broad shape maximises the surface area for absorption of sunlight and carbon dioxide.

The epidermal cells have no chloroplasts, so light passes through and reaches the palisade cells. They secrete a waxy cuticle, which reduces the evaporation of water from the upper surface of the leaf.

Most photosynthesis takes place in the palisade layer, as this is where the cells contain most chloroplasts. The position of this layer close to the surface of the leaf enables sunlight to reach it easily. The tall, narrow cells mean light has to pass through only three cell walls to reach the chloroplasts. The cells are tightly packed to capture the maximum sunlight.

Vascular bundles contain xylem vessels and phloem sieve tubes. Xylem helps to support the leaf, holding it out flat so it can absorb sunlight. Xylem vessels bring water to the leaf, which is necessary to keep cells turgid as well as being a reactant in photosynthesis. Phloem sieve tubes take away assimilates such as sucrose, made by photosynthesis.

The lower epidermal cells have no chloroplasts. They may secrete a waxy cuticle to cut down water loss, but this is usually thinner than on the upper surface.

Some photosynthesis takes place in the spongy layer, as there are some chloroplasts in these cells. The large air spaces between them allow easy and rapid diffusion of carbon dioxide from the stomata to the chloroplasts in the palisade layer and the spongy layer.

Two guard cells surround each stoma. The stomata allow rapid diffusion of carbon dioxide into the leaf for photosynthesis, and they connect with the air spaces inside the spongy layer. When the guard cells are turgid, they bend outwards and leave a space between them — the stomatal pore. When they are flaccid, they are less bent and the pore between them is closed.

Figure 13.10 Leaf structure in a C3 plant

In a C4 plant, rubisco and RuBP are kept away from the air spaces inside the leaf, so they do not come into contact with oxygen. The rubisco and RuBP are inside the chloroplasts of the **bundle sheath cells** (Figure 13.11). They are separated from the air spaces by a ring of **mesophyll cells**. These also contain chloroplasts, where the light dependent reactions of photosynthesis take place.

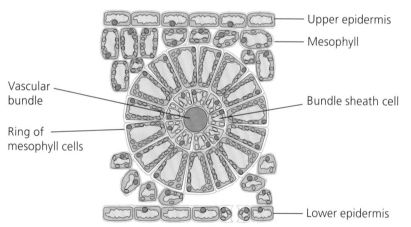

Upper epidermis
Mesophyll
Vascular bundle
Bundle sheath cell
Ring of mesophyll cells
Lower epidermis

Figure 13.11 Section through a leaf of a C4 plant

Carbon fixation happens like this:
- In the mesophyll cells, carbon dioxide combines with a substance called PEP to form a 4-carbon compound.
- The 4-carbon compound moves into the bundle sheath cells.
- The 4-carbon compound breaks down and releases carbon dioxide.
- The rubisco in the bundle sheath cells catalyses the reaction of the carbon dioxide with the RuBP.
- The Calvin cycle then proceeds as normal, inside the bundle sheath cells.

The enzymes involved in photosynthesis in maize and sorghum have **higher optimum temperatures** than in plants that are not adapted for growth in hot climates.

14 Homeostasis

Homeostasis in mammals

Negative feedback Revised

Homeostasis is the maintenance of a relatively constant internal environment in the body. Mammals are able to control many properties of their internal environment, including their internal temperature, glucose concentration and water potential. This is done using **negative feedback**. Negative feedback involves:

- the detection of changes in the internal environment or the external environment (stimuli) by receptors
- the passing of information about these changes to a central control area
- the dissemination of this information to different parts of the body, by coordination systems
- a response to the information by effectors (muscles and glands) to bring the changed feature of the internal environment back towards its set point

In negative feedback, a change in one direction — for example, an increase in blood glucose concentration — brings about a response by effectors that reverses the change.

Two coordination systems are involved in homeostasis in mammals:

- **The nervous system**. This is made up of the brain, spinal cord and nerves, which all contain specialised cells called neurones. Information is passed from one part of the body to another in the form of electrical impulses that pass very rapidly along neurones. You can read more about the nervous system on pp. 128–131.
- **The endocrine system**. This is made up a number of endocrine glands, which secrete hormones into the blood. Information is transferred as the hormones are distributed around the body in the blood.

> **Homeostasis** is the maintenance of a relatively constant environment for cells.
>
> **Negative feedback** is a mechanism by which a change in a particular factor brings about an action that reverses the change.

Thermoregulation Revised

Mammals and birds are able to maintain a fairly constant core (internal) temperature. In humans, the set point for core temperature is about 37 °C.

The receptors that detect changes in core temperature are called **thermoreceptors**, and are found in the hypothalamus in the brain. Thermoreceptors in the skin detect changes in the external temperature.

If these receptors detect a temperature rise:

- Impulses are sent along motor neurones to arterioles in the skin, causing the muscle in their walls to relax. This causes the arterioles to widen (vasodilation), allowing more blood to flow to capillaries at the skin surface. This allows more heat to be lost from the blood to the environment.
- Impulses are also sent along motor neurones to the sweat glands in the skin, causing them to secrete more sweat. The water in the sweat evaporates, taking heat from the skin in order to do this, and therefore cooling the body.

> **Now test yourself**
>
> 1 Explain the advantages of maintaining a constant core temperature.
>
> **Answer on p.204**
>
> 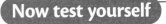 Tested

> **Typical mistake**
>
> Note that the arterioles simply get wider; they do not move up and down in the skin.

If the receptors detect a temperature fall:

- Impulses are sent along motor neurones to arterioles in the skin, causing the muscle in their walls to contract. This causes the arterioles to become narrower (vasoconstriction), allowing less blood to flow to capillaries at the skin surface. This allows less heat to be lost from the blood to the environment.

- Impulses are sent along sensory neurones to skeletal muscles, causing them to contract and relax very rapidly (shivering), generating heat that is transferred to the blood.

- In mammals other than humans, erector muscles are stimulated to contract, raising the hairs on end and trapping a layer of air between the hairs and the skin. Air is a good insulator, so this reduces heat loss from the body.

You can see that both of these sets of responses will help to bring back the core temperature to its set point. This control mechanism is a good example of negative feedback.

Control of blood glucose concentration

Blood glucose concentration should remain at a fairly constant value of about 100 mg glucose per 100 cm^3 of blood.

- If blood glucose concentration falls well below this level, the person is said to be hypoglycaemic. Cells do not have enough glucose to carry out respiration, and so metabolic reactions may not be able to take place and the cells cannot function normally. This is especially so for cells such as brain cells, which can only use glucose and not other respiratory substrates. The person may become unconscious and various tissues can be damaged.

- If blood glucose concentration rises well above this level, the person is said to be hyperglycaemic. The high glucose concentration decreases the water potential of the blood and tissue fluid, so that water moves out of cells down a water potential gradient. Again, unconsciousness can result.

Several hormones are involved in the control of blood glucose concentration by **negative feedback**. They include **insulin** and **glucagon**.

Both of these are small proteins. They are secreted by patches of tissue called **islets of Langerhans** in the pancreas (Figure 14.1). Insulin is secreted by β cells. Glucagon is secreted by α cells.

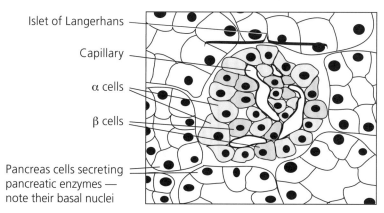

Figure 14.1 The histology of the pancreas

The β cells sense when blood glucose concentration rises too high. They respond by secreting greater quantities of insulin into the blood. The insulin has several effects, including:

- causing muscle and adipose tissue cells (fat cells) to absorb more glucose from the blood
- causing liver cells to convert glucose to glycogen for storage

> **Typical mistake**
>
> Note that it is the pancreas itself that senses changes in blood glucose concentration. Many students think it is the hypothalamus, but this is not involved in blood glucose regulation.

> **Typical mistake**
>
> Insulin is not an enzyme, and it does not convert glucose to glycogen. Insulin activates enzymes in the liver that cause this to take place.

These effects cause the blood glucose concentration to fall.

The α cells sense when blood glucose concentration falls too low. They respond by secreting greater quantities of glucagon into the blood. This has several effects, including:

- causing liver cells to break down glycogen to glucose, and releasing it into the blood
- causing liver cells to produce glucose from other substances such as amino acids or lipids

These effects cause blood glucose concentration to rise.

Cyclic AMP

Glucagon has its effect on liver cells by causing the cell to produce a substance called **cyclic AMP**. Cyclic AMP is an example of a **second messenger**. (The hormone that causes its activation is the 'first messenger'.) Cyclic AMP then triggers a series of reactions in the cell that results in the production of an enzyme that causes the conversion of glycogen to glucose. This is shown in Figure 14.2.

Revision activity

- In pencil, draw a flow diagram to show how temperature is regulated. Then, using the same diagram, erase and rewrite words to show how blood glucose concentration is regulated.

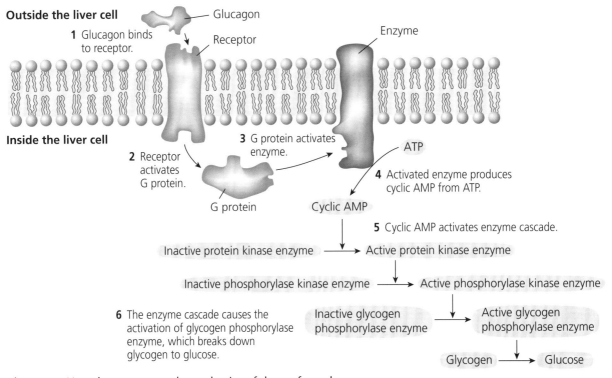

Figure 14.2 How glucagon causes the production of glucose from glycogen

This series of reactions is known as an **enzyme cascade**. Because each enzyme molecule that is activated can catalyse thousands more reactions, the number of activated enzymes increases at each step.

The hormone **adrenaline** has a similar effect. Adrenaline is secreted by the adrenal glands in a 'fight or flight' situation. It is transported in the blood to liver cells, where it binds to a different set of receptors on their cell surface membranes, activating the same enzyme cascade.

The interaction of the hormone with the liver cell is an example of **cell signalling**. This type of cell signalling involves:

- interaction between the hormone and a receptor on the cell surface membrane
- formation of cyclic AMP, which binds to kinase proteins
- an enzyme cascade involving activation of enzymes by phosphorylation to amplify the signal

Now test yourself

2 The enzyme cascade is said to 'amplify' the signal from glucagon. Suggest what this means.

Answer on p.204

Tested

Cell signalling is the transmission of information between one cell and another, by means of chemicals that bind with receptors on the cell surface membrane, or within the cell.

Dip sticks for measuring glucose concentration

Diabetes is a disease in which blood glucose concentration is not effectively controlled. People with diabetes need to measure their blood glucose concentration at regular intervals, so that they can control it by diet or by injecting insulin. Urine can also be tested for the presence of glucose. Normally, there should be no glucose in urine; its presence indicates diabetes.

Glucose concentration in a liquid, such as blood or urine, can be measured using dip sticks containing immobilised enzymes (Figure 14.3).

● A small pad at one end of the dip stick contains immobilised **glucose oxidase** and **peroxidase**. It also contains potassium iodide chromogen.
● When the pad is in contact with glucose, the glucose oxidase converts the glucose to gluconic acid and hydrogen peroxide.
● The peroxidase then catalyses a reaction of the hydrogen peroxide with the potassium iodide chromogen. Different colours are produced according to the quantity of hydrogen peroxide formed, which in turn depends on the original concentration of glucose in the liquid.

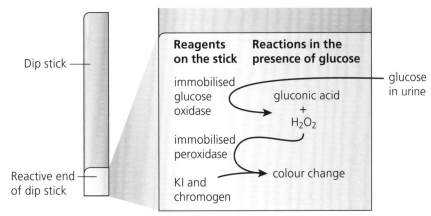

Figure 14.3 How a glucose dip stick works

Biosensors for measuring glucose concentration

A biosensor for glucose contains immobilised glucose oxidase. When in contact with glucose, the enzyme converts the glucose to gluconic acid. During this reaction, an enzyme cofactor is reduced. The reactions cause a flow of electrons, producing a tiny current that is measured and gives a digital readout.

Now test yourself

3 Suggest advantages of using a biosensor rather than a dip stick for measuring glucose concentration.

Answer on p.204

Tested

The roles of the kidneys in homeostasis

The kidneys help to maintain a constant internal environment by:
● excreting waste products, particularly the nitrogenous waste product urea
● helping to control the quantity of water in the body fluids (osmoregulation)

Excretion   Revised

Excretion is the removal of waste products generated by metabolic reactions inside body cells. Some of these products are toxic, while others are simply in excess of requirements.

Mammals excrete **urea**. Excess amino acids cannot be stored in the body. In the liver, they are converted to urea, $CO(NH_2)_2$, and a keto acid. This is known as **deamination** (Figure 14.4). The keto acid can be respired to provide energy, or converted to fat for storage.

Typical mistake

Remember that deamination takes place in the liver, not in the kidneys.

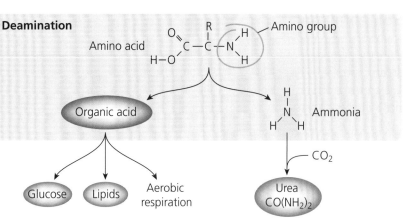

Figure 14.4 Formation of urea

The urea dissolves in the blood plasma and is removed and excreted by the kidneys. Urea contains nitrogen, and so is known as a nitrogenous excretory product.

The structure and histology of the kidneys

Revised

Each kidney is supplied with oxygenated blood through a renal artery. Blood is removed in the renal vein. A tube called the **ureter** takes urine from the kidney to the bladder (Figure 14.5).

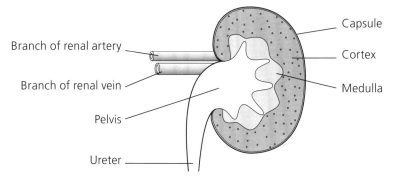

Figure 14.5 Section through a kidney and its associated blood vessels

Each kidney contains thousands of microscopic tubes called **nephrons** (Figure 14.6). The beginning of each nephron is a cup-shaped structure called a **renal capsule (Bowman's capsule)**. This is in the **cortex** of the kidney. The tube leads from the renal capsule down into the kidney **medulla**, then loops back into the cortex before finally running back down through the medulla into the **pelvis** of the kidney, where it joins the ureter.

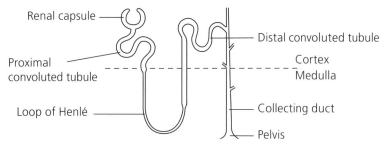

Figure 14.6 Structure of a nephron

Each nephron has a network of blood vessels associated with it. Blood arrives in the afferent arteriole (from the renal artery), and is delivered to a network of capillaries, called a glomerulus, in the cup of the renal capsule (Figure 14.7). Blood leaves the glomerulus in the efferent arteriole, which is narrower than the afferent arteriole. This leads to another network of capillaries that wraps around the nephron, before delivering the blood to a branch of the renal vein.

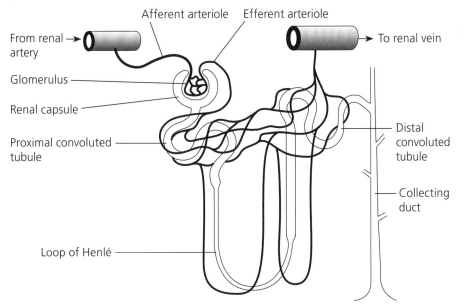

Figure 14.7 The blood vessels associated with a nephron

Figures 14.8 and 14.9 show the histology of the kidney. Histology is the structure of tissues.

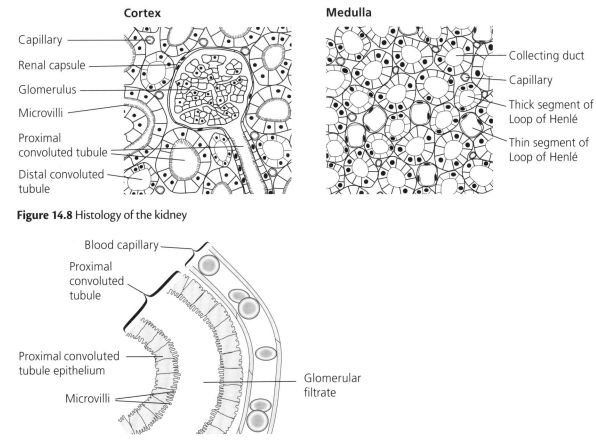

Figure 14.8 Histology of the kidney

Figure 14.9 Longitudinal section of part of a proximal convoluted tubule

How urine is produced in a nephron

Revised

Ultrafiltration

The blood in a glomerulus is separated from the space inside the renal capsule by:

- the capillary wall (**endothelium**) which is one cell thick and has pores in it
- the **basement membrane** of the wall of the renal capsule
- the layer of cells making up the wall of the renal capsule, called **podocytes**; these cells have slits between them

> **Ultrafiltration** is the separation of large solute molecules from small ones, by passing blood plasma through the basement membrane of the renal capsule.

The blood in a glomerulus is at a relatively high pressure, because the efferent arteriole is narrower than the afferent arteriole. This forces molecules from the blood through these three structures, into the renal capsule. The pores in the capillary endothelium and the slits between the podocytes will let all molecules through, but the basement membrane acts as a filter and will only let small molecules pass through.

Substances that can pass through include water, glucose, inorganic ions such as Na^+, K^+ and Cl^-, and urea.

Substances that cannot pass through include red and white blood cells and plasma proteins (such as albumen and fibrinogen).

The liquid that seeps through into the renal capsule is called glomerular filtrate (Table 14.1).

Table 14.1 Comparison of the composition of blood and glomerular filtrate

Component	Blood	Glomerular filtrate
Cells	Contains red cells, white cells and platelets	No cells
Water/g dm^{-3}	900	900
Inorganic ions (including Na^+, K^+ and Cl^-)/g dm^{-3}	7	7
Plasma proteins/g dm^{-3}	45	0
Glucose/g dm^{-3}	1	1
Urea/g dm^{-3}	0.3	0.3

Selective reabsorption in the proximal convoluted tubule

Some of the substances that are filtered into the renal capsule need to be retained by the body. These include:

- much of the water
- all of the glucose
- some of the inorganic ions

> **Selective reabsorption** is the removal of particular substances from the glomerular filtrate, and their return to the blood.

These substances are therefore taken back into the blood through the walls of the proximal convoluted tubule. This is called **selective reabsorption** (Figure 14.10).

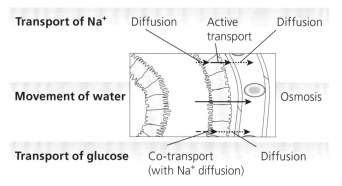

Figure 14.10 Selective reabsorption in the proximal convoluted tubule

The cells in the walls of the tubule have many mitochondria, to provide ATP for active transport. Their surfaces facing the lumen of the tubule have a large surface area provided by microvilli.

- **Active transport** is used to move Na^+ out of the outer surface of a cell in the wall of the proximal convoluted tubule, into the blood.
- This lowers the concentration of Na^+ inside the cell, so that Na^+ ions diffuse *into* the cell from the fluid inside the tubule. The Na^+ ions diffuse through protein transporters in the cell surface membrane of the cell.

- As the Na^+ ions diffuse through these transporter proteins, they carry glucose molecules with them. This is called **co-transport**. The glucose molecules move through the cell and diffuse into the blood.
- The movement of Na^+ and glucose into the blood decreases the water potential in the blood. Water therefore moves by **osmosis** from the fluid inside the tubule, down a water potential gradient through the cells making up the wall of the tubule and into the blood.

As a result, the fluid inside the nephron now has:

- no glucose
- a lower concentration of Na^+ than the filtrate originally had
- less water than the filtrate originally had

About 50% of the urea is also reabsorbed in the proximal convoluted tubule.

The loop of Henlé

Some, but not all, nephrons have long loops of Henlé that dip down into the medulla and then back up into the cortex. The function of the loop of Henlé is to build up a high concentration of Na^+ and Cl^- in the tissues of the medulla. This allows highly concentrated urine to be produced. Note that the loop of Henlé itself does not produce highly concentrated urine.

As fluid flows down the descending limb of the loop of Henlé, water moves out of it by osmosis. By the time the fluid reaches the bottom of the loop, it has a much lower water potential than at the top of the loop. As it flows up the ascending limb, Na^+ and Cl^- move out of the fluid into the surrounding tissues, first by diffusion and then by active transport.

This creates a low water potential in the tissues of the medulla. The longer the loop, the lower the water potential that can be produced.

The distal convoluted tubule and collecting duct

The fluid inside the tubule as it leaves the loop of Henlé and moves into the collecting duct has lost a little more water and more Na^+ than it had when it entered the loop. Because more water has been lost, the concentration of urea has increased.

Now, in the distal convoluted tubule, Na^+ is actively transported out of the fluid.

The fluid then flows through the collecting duct. This passes through the medulla, where you have seen that a low water potential has been produced by the loop of Henlé. As the fluid continues to flow through the collecting duct, water moves down the water potential gradient from the collecting duct and into the tissues of the medulla. This further increases the concentration of urea in the tubule. The fluid that finally leaves the collecting duct and flows into the ureter is **urine**.

Now test yourself

4 Although urea is not added to the liquid inside the proximal convoluted tubule, its concentration increases. Why is this?

Answer on p.204

Tested

Revision activity

- Construct a flow diagram showing what happens to glomerular filtrate as it passes through a nephron.

Osmoregulation

Revised

Osmoregulation is the control of the water content of body fluids. It is part of **homeostasis**, the maintenance of a constant internal environment. It is important that cells are surrounded by tissue fluid of a similar water potential to their own contents, to avoid too much water loss or gain, which could disrupt metabolism.

You have seen that water is lost from the fluid inside a nephron as it flows through the collecting duct. The permeability of the walls of the distal convoluted tubule and collecting duct can be varied.

- If the walls are permeable to water, then much water can move out of the tubule and the urine becomes concentrated. The water is taken back into the blood and retained in the body.

Now test yourself

5 Explain, using water potential terminology, what would happen to a red blood cell if the water potential of the blood plasma dropped too low.

Answer on p.204

Tested

● If the walls are made impermeable to water, little water can move out of the tubule and the urine remains dilute. A lot of water is removed from the body.

ADH

ADH is **antidiuretic hormone**. It is secreted from the **posterior pituitary gland** into the blood.

Osmoreceptor cells in the **hypothalamus** sense when the water potential of the blood is too low (that is, it has too little water in it). The osmoreceptor cells are neurones (nerve cells). They produce ADH, which moves along their axons and into the posterior pituitary gland from where it is secreted into the blood (Figure 14.11).

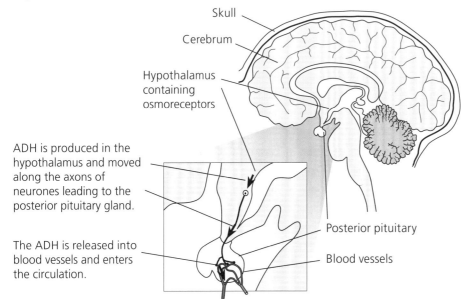

Skull

Cerebrum

Hypothalamus containing osmoreceptors

ADH is produced in the hypothalamus and moved along the axons of neurones leading to the posterior pituitary gland.

The ADH is released into blood vessels and enters the circulation.

Posterior pituitary

Blood vessels

Figure 14.11 Osmoreceptors and the secretion of ADH

The ADH travels in solution in the blood plasma. When it reaches the walls of the collecting duct, it makes them permeable to water. Water is therefore reabsorbed from the fluid in the collecting duct and small volumes of concentrated urine are produced.

The osmoreceptor cells also sense when the water potential of the blood is too high (that is, it has too much water in it), and less ADH is secreted. The collecting duct walls therefore become less permeable to water and less is reabsorbed into the blood. Large volumes of dilute urine are produced.

Negative feedback

The mechanism for controlling the water content of the body, using ADH, is an example of **negative feedback**.

When the water potential of the blood rises too high or falls too low, this is sensed by **receptor** cells. They cause an action to be taken by **effectors**, which cause the water potential to be moved back towards the correct value.

In this case, the receptors are the osmoreceptor cells in the hypothalamus, and the effectors are their endings in the anterior pituitary gland that secrete ADH.

Urine analysis

We have seen on p. 121 that urine can be tested for glucose; its presence indicates diabetes. Urine can also be tested for protein and ketones. The presence of protein indicates kidney damage or an infection, which is allowing proteins from the blood to enter the filtrate. It can also be an indication of high blood pressure. Ketones are produced by the metabolism of fatty acids, and ketones in urine indicates that they are in higher than normal concentrations in

> **Revision activity**
>
> ● Use the diagram that you constructed to show how blood glucose concentration and temperature are controlled, but now adapt it to show how the water potential of the blood is controlled.

the blood, suggesting that body cells do not have sufficient supplies of glucose and are therefore using fatty acids instead. This may indicate diabetes, but can also indicate that the person is not eating sufficient carbohydrate.

Homeostasis in plants

Plants are not able to control their internal environment in the way that mammals do. However, they are able to regulate water loss to some extent, and also respond to changes in environmental conditions, by opening or closing their stomata.

A stoma is a pore in the epidermis of a plant that is surrounded by a pair of guard cells (Figure 14.12). Most stomata are in the lower surfaces of leaves.

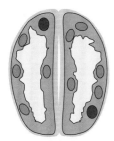
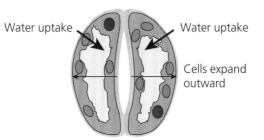

Water uptake   Water uptake

Cells expand outward

Figure 14.12 A stoma and guard cells

The cell walls of the guard cells are thicker nearer the stomatal aperture than on their outer edges. When they take up water, their expansion causes the stoma to open.

- Proton pumps in the guard cell membranes move protons (hydrogen ions) out of the cells by active transport, which produces a negative electrical potential inside the cell.
- This causes potassium ion channels in the cell surface membrane to open.
- Potassium ions diffuse into the cell through the open channels, down an electrochemical gradient.
- This decreases the water potential of the cell, so water moves into the cell by osmosis, down a water potential gradient.
- This increases the volume of the cell, causing it to swell and curve away from its partner, opening the stoma.
- In the presence of very high temperatures, or when the plant is short of water, a plant hormone (plant growth regulator) called **abscisic acid**, **ABA**, is produced.
- It is thought that ABA binds with receptors on the cell surface membranes of the guard cells, which inactivates the proton pumps.
- ABA also stimulates the uptake of calcium ions into the cell.
- The calcium ions act as a second messenger, activating channel proteins that allow negatively charged ions (anions) such as chloride ions to move out of the cell by facilitated diffusion. The calcium ions also cause potassium ion channels to open, so that potassium ions can leave the cell.
- The exit of chloride ions and potassium ions increases the water potential of the cell, so water moves out down a water potential gradient by osmosis.
- The guard cells therefore return to their normal shape and close the stoma.

In many plant species, the opening and closing of stomata follows a natural diurnal (daily) rhythm. Stomata tend to close at night and open in the daytime, when light levels are high, so that carbon dioxide can diffuse into the leaf and photosynthesis can take place. However, this rhythm is altered in response to environmental conditions such as high temperatures or low availability of water, which can cause stomata to close in the daytime.

15 Control and coordination

In multicellular organisms, such as plants and animals, it is essential that cells can communicate with each other. This allows them to coordinate their activities appropriately. Organisms have specialised cells or molecules, called **receptors**, which are sensitive to changes in their internal or external environment. These trigger events in the organism that bring about coordinated responses to the environmental changes.

Control and coordination in animals

Neurones Revised

Neurones (nerve cells — Figure 15.1) are highly specialised cells that are adapted for the rapid transmission of electrical impulses, called **action potentials**, from one part of the body to another.

Motor neurone

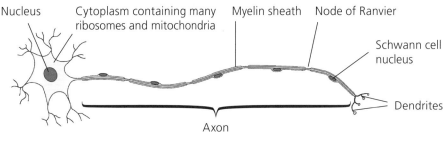

Sensory neurone   **Relay neurone**

Figure 15.1 The structure of neurones

Information picked up by a **receptor** is transmitted to the **central nervous system** (brain or spinal cord) as action potentials travelling along a **sensory neurone**. These neurones have their cell bodies in small swellings, called **ganglia**, just outside the **spinal cord**. The impulse may then be transmitted to a **relay neurone**, which lies entirely within the brain or spinal cord. The impulse is then transmitted to many other neurones, one of which may be a **motor neurone**. This has its cell body within the central nervous system, and a long axon, which carries the impulse all the way to an **effector** (a muscle or gland).

In some cases, the impulse is sent on to an effector before it reaches the 'conscious' areas of the brain. The response is therefore automatic, and does not involve any decision-making. This type of response is called a **reflex**, and the arrangement of neurones is called a **reflex arc** (Figure 15.2).

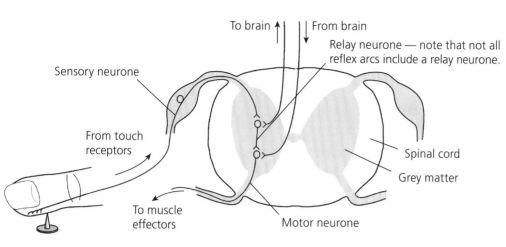

Figure 15.2 Arrangement of neurones in a reflex arc

The **myelin sheath** is made up of many layers of cell membranes of **Schwann cells**, which wrap themselves round and round the axon. This provides electrical insulation around the axon, which greatly speeds up the transmission of action potentials. Not all neurones are myelinated.

Resting potential

Neurones, like all cells, have sodium–potassium pumps in their cell surface membranes. However, in neurones these are especially active. They pump out sodium ions and bring in potassium ions by active transport. Three sodium ions are moved out of the cell for every two potassium ions that are moved in.

There are also other channels in the membrane that allow the passage of sodium and potassium ions. These are **voltage-gated channels**. The potential difference across the membrane determines whether they are open or closed. When a neurone is resting, quite a few potassium ion channels are open, so potassium ions are able to diffuse back out of the cell, down their concentration gradient.

As a result, the neurone has more positive ions outside it than inside it. This means there is a potential difference (a voltage) across the axon membrane. It has a charge of about **−70 mV** (millivolts) inside compared with outside. This is called the **resting potential**.

> The **resting potential** is a difference in electrical potential between the outside and inside of the cell surface membrane of a neurone; usually about −70 mV inside.

> **Expert tip**
>
> Always take care to write about sodium *ions* and potassium *ions*, not just 'sodium' and 'potassium'. Alternatively, you can use their symbols: Na$^+$ and K$^+$.

Generation of an action potential by receptors

Receptors are specialised cells that respond to stimuli by generating action potentials. Figure 15.3 shows an example of a receptor — a taste bud in the tongue.

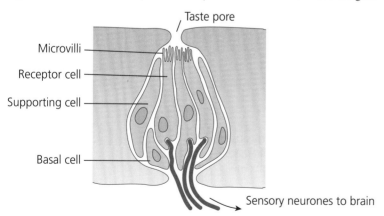

Figure 15.3 A taste bud

The cell surface membranes of the microvilli contain proteins that act as receptors for chemicals in food. When a particular chemical binds with this receptor, it causes sodium channels to open in the membrane. This allows sodium ions to diffuse into the cell down an electrochemical gradient and depolarise it. This quickly reverses the potential difference across the cell membrane, making it much less negative inside. The neurone is said to be **depolarised**. If this depolarisation is relatively small, then nothing further happens. However, if the depolarisation is great enough, then an **action potential** is triggered. The sodium ions keep on flooding in until the cell has actually become positive inside, reaching a potential of about +30 mV. The sodium ion channels then close.

This change in potential difference across the membrane causes a set of voltage-gated potassium ion channels to open. Potassium ions can now flood out of the axon, down their electrochemical gradient. This makes the charge inside the axon less positive. It quickly drops back down to a little below the value of the resting potential.

The voltage-gated potassium ion channels then close and the resting potential is restored (Figure 15.4).

The time taken for the axon to restore its resting potential after an action potential is called the **refractory period**. The axon is unable to generate another action potential until the refractory period is over. The refractory period therefore places an upper limit for the frequency of nerve impulses.

> **Depolarisation** is the reduction or reversal of the resting potential.

> **Expert tip**
> Note that all action potentials are the same size. Stronger stimuli do not generate larger action potentials. They do, however, generate more frequent action potentials.

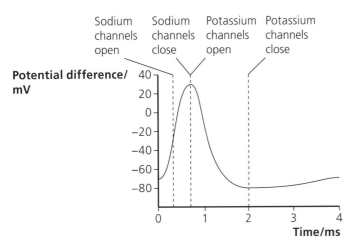

Figure 15.4 An action potential

> **Now test yourself**
> 1 Explain what causes the opening and closing of the K$^+$ channels shown in Figure 15.4.
>
> **Answer on p.204**
>
> Tested

Transmission of action potentials

An action potential or nerve impulse that is generated in one part of a neurone travels rapidly along its axon or dendron. This happens because the depolarisation of one part of the membrane sets up local circuits with the areas on either side of it. These cause depolarisation of these regions as well. The nerve impulse therefore sweeps along the axon (Figure 15.5).

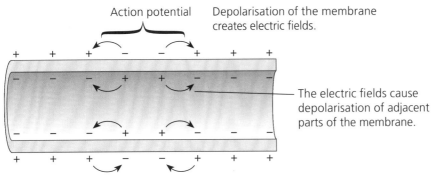

Figure 15.5 How a nerve impulse travels along a neurone

In a myelinated neurone, local circuits cannot be set up in the parts of the neurone where the myelin sheath is present. Instead, the nerve impulse 'jumps' from one **node of Ranvier** to the next. This is called **saltatory conduction**. This greatly increases the speed at which the action potential travels along the axon.

Synapses

Where two neurones meet, they do not actually touch. There is a small gap between them called a **synaptic cleft**. The membrane of the neurone just before the synapse is called the **presynaptic membrane**, and the one on the other side is the **postsynaptic membrane**. The whole structure is called a **synapse** (Figure 15.6).

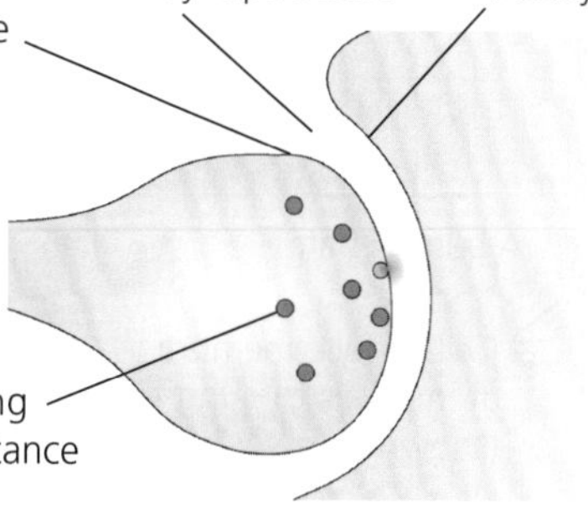

Presynaptic membrane   Synaptic cleft   Postsynaptic membrane

Vesicle containing transmitter substance

Figure 15.6 A synapse

Expert tip

Take care not to use the word 'synapse' when you mean 'synaptic cleft'.

- When an action potential arrives at the presynaptic membrane, it causes voltage-gated calcium ion channels to open.
- Calcium ions rapidly diffuse into the cytoplasm of the neurone, down their concentration gradient.
- The calcium ions affect tiny vesicles inside the neurone, which contain a **neurotransmitter** such as **acetylcholine**. These vesicles move towards the presynaptic membrane and fuse with it, releasing their contents into the cleft.
- The neurotransmitter diffuses across the cleft and slots into receptor molecules in the postsynaptic membrane.
- This causes sodium ion channels to open, so sodium ions flood into the cytoplasm of the neurone, depolarising it.
- This depolarisation sets up an action potential in the postsynaptic neurone.

Functions of synapses in the body

- Although action potentials are able to travel in either direction along a neurone, synapses ensure that action potentials can only travel one way.
- One neurone may have synapses with many other neurones. This allows interconnection of nerve pathways from different parts of the body.
- Synapses allow for a wide variety of responses by effectors. For example, a motor neurone may need to receive transmitter substances from many different neurones forming synapses with it before an action potential is generated in it. Or some of these neurones may produce transmitter substances that actually *reduce* the chance of an action potential being produced. The balance between the signals from all these different synapses determines whether or not an action potential is produced in the motor neurone, and therefore whether or not a particular effector takes action.

Now test yourself

2 Explain why a nerve impulse can only travel in one direction across a synapse.

Answer on p.204

Tested

Muscles   Revised

Muscles are effectors that are able to cause movement. The muscles that are attached to the skeleton are called skeletal or **striated muscles**.

Striated muscle tissue is made up of 'cells' called **muscle fibres**. These are not normal cells, because each one contains several nuclei.

Each muscle fibre is made up of many parallel **myofibrils** (Figure 15.7). The myofibrils contain several different proteins, including **actin** and **myosin**. The muscle fibres also contain many mitochondria, which supply the ATP needed for muscle contraction. The endoplasmic reticulum of the muscle fibre is very specialised, and forms **T-tubules** that run from the cell surface membrane right into the centre of the fibre. Branching from these T-tubules are cisternae in which calcium ions accumulate.

Part of a muscle fibre
Muscle fibres are very specialised cells containing many nuclei and are up to several centimetres long.

Nucleus

Myofibril
Myofibrils are collections of actin and myosin filaments.

Sarcomere

Myofibril

A band I band

Cytoplasm — contains myoglobin (for clarity organelles such as mitochondria and ER not shown)

Cell surface membrane — infolded at intervals forming t-tubules (shown in lower half only)

Detail of a myofibril at the edge of a muscle fibre

Cell surface membrane

t-tubule

Cisternum of ER containing Ca^{2+}

Mitochondrion

Cytoplasm containing myoglobin

Figure 15.7 The structure of striated muscle

Actin molecules form **thin filaments**, and bundles of myosin molecules forms **thick filaments**. These filaments are arranged in a very specific pattern within myofibrils, to form **sarcomeres** (Figure 15.8). Each myofibril contains many sarcomeres arranged in a continuous line.

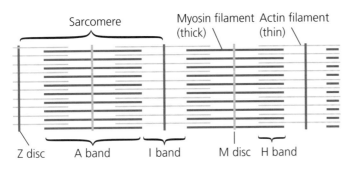

Sarcomere

Myosin filament (thick) Actin filament (thin)

Z disc   A band   I band   M disc   H band

Figure 15.8 Sarcomeres in a myofibril

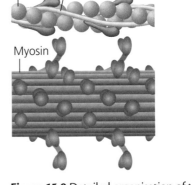

Actin   Tropomyosin   Troponin

Myosin

Figure 15.9 Detailed organisation of the proteins in part of a sarcomere

Two other proteins, called troponin and tropomyosin, are associated with the actin filaments (Figure 15.9). In a resting muscle, troponin covers sites on actin to which myosin would otherwise be able to bind.

Now test yourself

Tested

3 Use Figures 15.7–15.9 to explain the difference between a muscle fibre, a myofibril and a filament.

Answer on p.204

Muscle stimulation and contraction

Striated muscle is stimulated to contract when a nerve impulse arrives at a **neuromuscular junction**. This is like a synapse, but the postsynaptic membrane is the membrane of a muscle fibre rather than the membrane of a neurone. When a nerve impulse arrives, the following stages occur:

- The action potential spreads down the membranes of the T-tubules, which carry it deep into each myofibril.
- The arrival of the action potential opens calcium ion channels in these membranes, allowing calcium ions to diffuse out of the cisternae and into the cytoplasm of the myofibril.
- The calcium ions bind to the troponin molecules.
- This causes the troponin molecules to change shape, which in turn causes the tropomyosin molecules to move away from the binding sites they are covering.
- The myosin heads can now bind to the actin filaments, forming actin–myosin bridges.
- The myosin heads tilt, pushing the actin filaments along. The actin filaments slide between the myosin filaments, causing the Z discs to be pulled closer together and the sarcomere to get shorter.
- The myosin heads detach from the actin filament, flip back to their normal shape and reattach. They then tilt again, pushing the actin filament further along (Figure 15.10). This can be repeated many times a second.

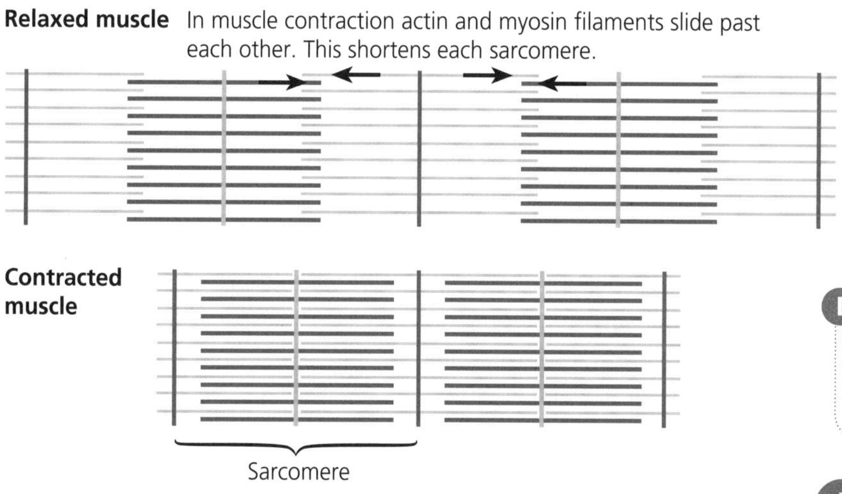

Relaxed muscle In muscle contraction actin and myosin filaments slide past each other. This shortens each sarcomere.

Contracted muscle

Sarcomere

Figure 15.10 How sliding filaments cause muscle contraction

The energy from this process comes from the hydrolysis of ATP. ATP binds to the myosin heads after they have tilted. The ATP is then hydrolysed to ADP and P_i. The energy that is released causes the myosin heads to tilt back to their original position, so that they can bind again with actin.

> **Expert tip**
>
> Note that the ATP is used to detach the myosin from the actin, not to make it bind or move.

> **Now test yourself**
>
> 4 Explain why muscle fibres contain many mitochondria.
>
> **Answer on p.204**
>
> Tested

The mammalian endocrine system

Revised

The endocrine system is made up of a number of **endocrine glands**. These are organs containing cells that secrete **hormones**. The hormones are secreted directly into the blood, not into a duct as with other types of gland (for example the salivary glands, which secrete saliva into the salivary ducts).

Hormones secreted by an endocrine gland are transported in solution in the blood plasma all over the body. Certain cells have receptors for these hormones in their cell surface membranes. These cells are the **target cells** for the hormones. For example, the target cells for ADH (p. 126) are the cells lining the collecting duct in a nephron. The target cells for insulin and glucagon include liver cells (p. 119).

Reproductive hormones in the menstrual cycle

In humans, the menstrual cycle is a repeating process of change in the **ovaries**, **oviducts** and **uterus**, which takes place approximately every 28 days from puberty to menopause. It is controlled by four hormones:

- two **steroid hormones** secreted by the ovaries — **oestrogen** and **progesterone**
- two **gonadotropic peptide hormones** secreted by the **anterior pituitary gland** — **FSH** (follicle stimulating hormone) and **LH** (luteinising hormone).

First half of the cycle

During the first half of each cycle, the dominant gonadotropin is FSH. This stimulates secondary follicles in the ovary to grow in size and number. The granulosa cells of the secondary follicles secrete oestrogen, which is the dominant steroid hormone in this stage of the cycle. This affects:

- the oviduct, causing the development of more, larger and more active cilia on the cells lining its walls. It also increases the secretion of more glycoprotein; these changes prepare the oviduct for the arrival of an oocyte, which will be moved along the oviduct by the cilia.
- the endometrium (lining of the uterus), causing the cells to divide to form a thicker layer, ready to receive an embryo if the egg is fertilised

Mid-point of the cycle

At day 13 or 14, a surge in LH causes the primary oocyte in a single ovarian follicle to complete meiosis I and continue to metaphase of meiosis II. It also causes the follicle to shed the secondary oocyte into the oviduct. This is ovulation.

Second half of the cycle

The LH surge also causes the granulosa cells lining the secondary follicles to change to luteal cells, and switch to secreting largely progesterone and less oestrogen.

Progesterone becomes the dominant steroid hormone from 5 days after ovulation (which is when a fertilised embryo would enter the uterus). Progesterone causes the cells in the endometrium to differentiate to form a thick, vascularised layer. If fertilisation does not occur, the corpus luteum begins to degenerate from day 23, so the secretion of progesterone also decreases. This causes the breakdown of the endometrium, leading to the start of menstruation on day 0 of the start of the next cycle (Figure 15.11).

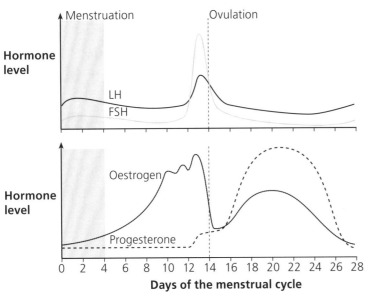

Figure 15.11 Changes in hormones during the menstrual cycle

Negative feedback in the menstrual cycle

High concentrations of oestrogen inhibit the secretion of FSH and LH by the anterior pituitary gland. This happens during the first half of the menstrual cycle, causing the levels of FSH and LH to fall. However, when oestrogen levels are very high, a surge of LH secretion occurs, which brings about ovulation.

Towards the end of the cycle, as oestrogen and progesterone levels fall, the inhibition of FSH and LH secretion is lifted, so the concentration of these two hormones begins to increase, leading to the start of a new cycle.

The contraceptive pill

Most contraceptive pills contain synthetic hormones similar to progesterone and oestrogen. These prevent the secretion of LH and FSH from the anterior pituitary gland. There is therefore no development of secondary follicles in the ovary (which is caused by FSH) and no ovulation (which is caused by a surge of LH).

> **Expert tip**
>
> Remember to say that FSH and LH are produced by the **anterior** pituitary gland. ADH is produced by the **posterior** pituitary gland.

> **Revision activity**
>
> ● Convert the written description of the menstrual cycle on pp. 134–135 to a flow diagram.

Table 15.1 Comparison of the mammalian nervous and endocrine systems

Feature	Nervous system	Endocrine system
Components	Made up of neurones in nerves, brain and spinal cord	Made up of secretory cells in endocrine glands
How information is transmitted	As electrical impulses in neurones	As chemicals (hormones) carried in blood plasma
Speed of transmission	Very fast	Slower
Duration of effect	Often very brief	Often longer term
Breadth of effect	Often affects only a few effector cells	Often affects several different target organs

Control and coordination in plants

As in animals, it is essential that cells in a plant can communicate with each other. This allows them to coordinate their activities appropriately, and allows the entire plant to respond to changes in its internal or external environment. As in animals, plant communication systems involve **receptors** and **effectors**. Plants have a system of electrical communication similar to that of an animal's nervous system, but 'action potentials' travel only very slowly and are very weak.

Venus fly trap Revised

The Venus fly trap (Figure 15.12) grows in soils that are low in nitrate. In order to obtain sufficient nitrogen for protein synthesis, the Venus fly trap digests insects and absorbs nutrients from them.

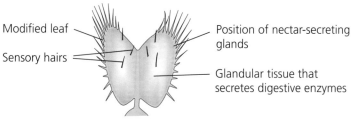

Modified leaf

Sensory hairs

Position of nectar-secreting glands

Glandular tissue that secretes digestive enzymes

Figure 15.12 Venus fly trap

Nectar attracts insects such as flies to the modified leaf. The insect's movements cause the sensory hairs to move, and this stimulus causes calcium ion channels in cells at the base of the hair to open. Calcium ions flow in, depolarising the cells. If two or more hairs are stimulated, or if the same hair is stimulated twice in quick succession, this depolarisation is great enough to trigger an action potential.

The action potential travels across the leaf, and causes the lobes to change into a convex shape, which makes them fold together and close the trap. The stiff spikes on the outer surface of the trap slot together, sealing the trap.

Hydrolytic enzymes are then secreted, and these digest the prey to soluble substances such as amino acids, which are absorbed into the leaf.

Plant growth regulators
Revised

Like animal hormones, plant growth regulators are chemicals that act on target cells, where they bind with receptors either on the cell surface membrane or inside the cell, and bring about changes. Unlike animal hormones, plant growth regulators are not made in specialised glands.

> **Plant growth regulators** are hormone-like substances that are produced in a plant and have effects on target cells that carry receptors for them.

Many different plant growth regulators have been discovered, but there is still much that we do not know about their actions. They often have different effects at different concentrations, in different parts of a plant, at different stages of its life cycle or when other hormones are present.

Auxins

Auxins are a group of plant hormones that are produced by cells in regions of cell division, or **meristems**. In a plant shoot, the apical bud (the bud at the tip of the growing shoot) contains a meristem. Auxin is constantly made here. Auxin molecules are then moved down the shoot from cell to cell, through special auxin transporter proteins in the cell surface membranes.

Now test yourself

5   Construct a table to compare plant growth regulators with animal hormones.

Answer on p.204

Tested

Auxin has many different effects in plants, one of which is to stimulate elongation of cells, which results in growth. It does this by stimulating proton pumps in cell surface membranes to remove protons from the cytoplasm and deposit them in the cell walls. This acidification of the cell walls loosens the bonds between the cellulose molecules and matrix, allowing the cell to elongate as it absorbs water.

Gibberellins

Gibberellin, also known as GA, is produced in the embryos of germinating seeds, including wheat and barley, after they have absorbed water. GA switches on several genes that encode enzymes that hydrolyse food reserves. (See also p. 153.) These include starch, protein and lipid that are stored in the endosperm of the seed. For example, amylase is produced, which hydrolyses starch to maltose. The soluble maltose is transported to the embryo and used as an energy source and a raw material for the production of new cells.

GA also stimulates stem elongation, by causing cell division and cell elongation. Some plants lack the last enzyme in the metabolic pathway by which gibberellin is synthesised, and they remain small. The gene for this enzyme has two alleles, **Le** and **le**. Allele **Le** is dominant and codes for the enzyme. Plants homozygous for **le** do not produce the enzyme. Applying gibberellin to these genetically dwarf plants makes them grow tall.

Now test yourself

6   Explain why applying gibberellin to plants with the genotype **LeLe** does not make them grow taller.

Answer on p.204

Tested

Abscisic acid
See p. 127 for the role of abscisic acid in stomatal closure.

16 Inherited change

Meiosis

A few cells in the human body — some of those in the testes and ovaries — are able to divide by a type of division called **meiosis**. Meiosis involves two divisions (not one as in mitosis), so four daughter cells are formed. Meiosis produces four new cells with:

- only half the number of chromosomes as the parent cell
- different combinations of alleles from each other and from the parent cell

In animals and flowering plants, meiosis produces **gametes**.

In a human, body cells are **diploid**, containing two complete sets of chromosomes. Meiosis produces gametes that are **haploid**, containing one complete set of chromosomes. This is necessary so that the cell formed at fertilisation (the zygote) is diploid.

Expert tip

Take great care to spell 'meiosis' correctly, so that it cannot be confused with 'mitosis'. You will not be given benefit of doubt if the examiner is not sure which one you mean.

Meiosis is a type of cell division in which a diploid cell divides to form four haploid cells, in which different combinations of alleles are present.

Typical mistake

Do not say that meiosis takes place 'in' gametes. It takes place in cells that produce gametes.

Gametogenesis in mammals
Revised

Gametogenesis is the production of haploid gametes from diploid somatic (body) cells.

In mammals, the male gametes are spermatozoa (sperm) and the female gametes are ova. Spermatogenesis takes place in the testes, and oogenesis in the ovaries.

Spermatogenesis

Diploid spermatogonia at the edge of the seminiferous tubule (Figure 16.1) undergo mitosis and then meiosis to produce haploid spermatids. As meiosis proceeds, the cells move towards the centre of the tubule. The whole process takes about 64 days (Figure 16.2).

n stands for the number of chromosomes in a complete set. A cell with one set (n) is haploid. A cell with two sets (2n) is diploid.

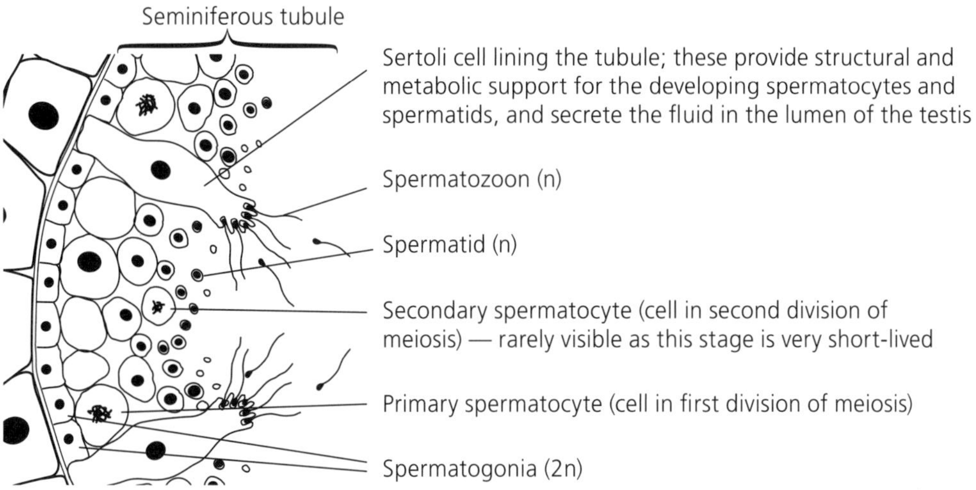

Seminiferous tubule

Sertoli cell lining the tubule; these provide structural and metabolic support for the developing spermatocytes and spermatids, and secrete the fluid in the lumen of the testis

Spermatozoon (n)

Spermatid (n)

Secondary spermatocyte (cell in second division of meiosis) — rarely visible as this stage is very short-lived

Primary spermatocyte (cell in first division of meiosis)

Spermatogonia (2n)

Figure 16.1 Histology of the testis

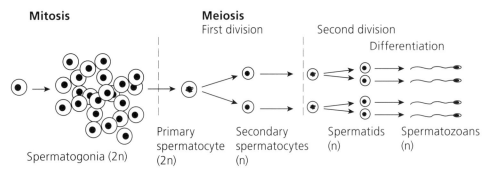

Figure 16.2 Sequence of spermatogenesis

Oogenesis

Diploid oogonia divide by mitosis and then meiosis to produce haploid secondary oocytes. At birth, the oogonia have already become primary oocytes, in the first division of meiosis. The primary oocytes are inside primordial follicles (Figure 16.3), in which the oocyte is surrounded by a layer of granulosa cells. The primordial follicle remains in this state for many years.

At puberty, some of the primordial follicles develop into primary follicles, in which the primary oocyte grows larger and develops a coat called the zona pellucida. The primary follicle then develops into a secondary follicle, containing extra layers of granulosa cells, which are surrounded by a theca.

From puberty onwards, some of the primary follicles develop into ovarian follicles. Just before ovulation, the primary oocyte completes the first division of meiosis to produce a secondary oocyte and a tiny polar body. At ovulation, the secondary oocyte is in metaphase of the second division of meiosis (Figure 16.4). Meiosis only completes after the egg is fertilised.

After ovulation, the remains of the ovarian follicle develop into a corpus luteum.

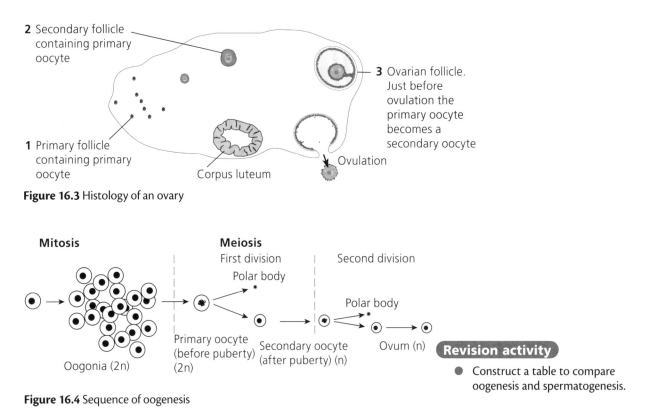

Figure 16.3 Histology of an ovary

Figure 16.4 Sequence of oogenesis

Revision activity

● Construct a table to compare oogenesis and spermatogenesis.

Gametogenesis in flowering plants

In flowering plants, the male gametes are nuclei inside pollen grains, which are made in the anthers of a flower. The female gametes are nuclei inside embryo sacs, which are made in the ovules inside the ovaries of a flower.

> **Typical mistake**
>
> Pollen grains are not male gametes. They *contain* the male gametes.

Production of male gametes

Inside the anthers, **pollen mother cells** divide by meiosis to form four haploid cells. The nuclei in each of these haploid cells divide by mitosis to produce two haploid nuclei in each cell (Figure 16.5). These cells mature into **pollen grains**. One of the nuclei is the **male gamete nucleus**, which can fuse with a female nucleus in an embryo sac to produce a diploid zygote. This zygote will grow into an embryo plant.

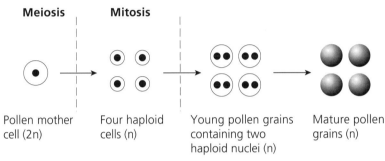

Figure 16.5 Production of male gametes in a flowering plant

Production of female gametes

Inside the ovary of a flower are one or more **ovules**. Inside the ovule, a diploid **embyro sac mother cell** divides by meiosis to form four haploid cells. Only one of these continues to develop (Figure 16.6). Its haploid nucleus divides by mitosis three times, forming a total of four haploid nuclei inside a structure called an **embryo sac**. One of these nuclei is the **female gamete nucleus** (in the egg cell) which can be fertilised by a male gamete nucleus from a pollen grain.

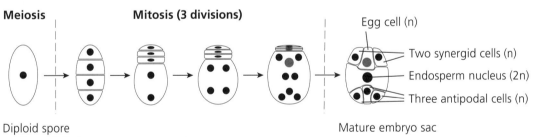

Figure 16.6 Production of a female gamete in a flowering plant

Behaviour of chromosomes during meiosis

Before meiosis begins, DNA replication takes place exactly as it does before mitosis (p. 40). However, in the early stages of meiosis **homologous chromosomes** (the two 'matching' chromosomes in a nucleus) pair up. Figure 16.7 shows how meiosis takes place.

> **Homologous chromosomes** are chromosomes that carry the same genes (though not necessarily the same alleles of the genes) in the same position.

Prophase I

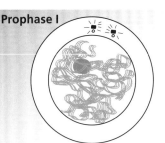
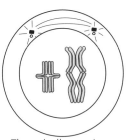

- Centrioles divide and are involved in spindle formation.
- Chromosomes beginning to condense.

- The spindle continues to form.
- Chromosomes pair and are now visible as bivalents.

Metaphase I

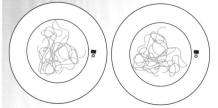

- The nuclear envelope disappears.
- The bivalents are arranged on the equator.

Anaphase I

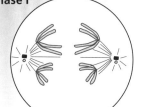

- Homologous chromosomes separate to opposite poles, pulled by spindle fibres.

Telophase I and cytokinesis

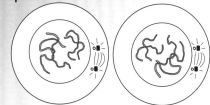

- Chromosomes partly unwind
- Spindle disappears and the nuclear envelopes form
- The cell divides

Prophase II

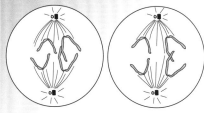

- Centrioles divide and spindles form
- Chromosomes condense

Metaphase II

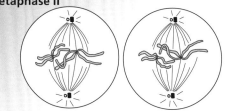

- Individual chromosomes line up on equator.

Anaphase II

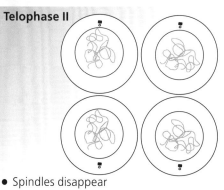

- Centromeres split
- Chromatids move apart, pulled by spindle fibres

Telophase II

- Spindles disappear
- Nuclear envelopes appear
- Chromosomes unwind

Figure 16.7 Meiosis

How meiosis causes genetic variation

Revised

Meiosis produces genetic variation. This is done by

- crossing over
- independent assortment

Further genetic variation is produced when gametes produced by meiosis fuse together to produce a zygote.

Crossing over

Each chromosome in a homologous pair carries genes for the same characteristics at the same locus. The alleles of the genes on the two chromosomes may be the same or different.

During prophase of meiosis I, as the two homologous chromosomes lie side by side, their chromatids form links called **chiasmata** (singular: chiasma) with each other. When they move apart, a piece of chromatid from one chromosome may swap places with a piece from the other chromosome. This is called **crossing over**. It results in each chromosome having different combinations of alleles than it did before (Figure 16.8).

Homologous pair of chromosomes

Chiasma

Pieces of each chromatid have swapped places, resulting in new combinations of alleles

Figure 16.8 Crossing over in meiosis

Independent assortment

Another feature of meiosis that results in the shuffling of alleles — and therefore genetic variation — is **independent assortment**. During the first division of meiosis, the pairs of homologous chromosomes line up on the equator before being pulled to opposite ends of the cell. Each pair behaves independently from every other pair, so there are many different combinations that can end up together. Figure 16.9 shows the different combinations you can get with just two pairs of chromosomes. In a human there are 23 pairs, so there is a huge number of different possibilities.

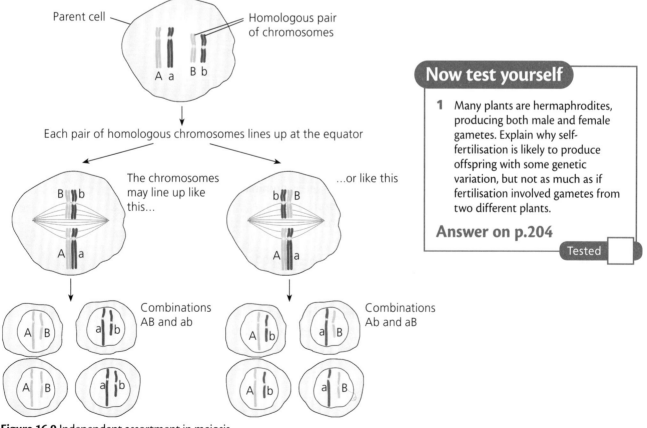

Parent cell

Homologous pair of chromosomes

A a   B b

Each pair of homologous chromosomes lines up at the equator

The chromosomes may line up like this...

B b

A a

...or like this

b B

A a

Combinations AB and ab

A B   a b

A B   a b

Combinations Ab and aB

A b   a B

A b   a B

Figure 16.9 Independent assortment in meiosis

Now test yourself

1 Many plants are hermaphrodites, producing both male and female gametes. Explain why self-fertilisation is likely to produce offspring with some genetic variation, but not as much as if fertilisation involved gametes from two different plants.

Answer on p.204

Tested

Genetics

Genes are passed from parents to offspring inside the nuclei of gametes. During sexual reproduction, gametes fuse to produce a zygote, which contains one set of chromosomes from each parent. The study of the passage of genes from parent to offspring is called genetics.

Genes and alleles

The two chromosomes in a diploid cell that are similar (e.g. the two chromosome 1s) are said to be **homologous**. They each contain the same genes in the same position, known as the **locus** of that gene. This means that there are two copies of each gene in a diploid cell.

Genes often come in different forms. For example, the gene for a protein that forms a chloride transporter channel in cell surface membranes, called the CFTR protein, has a normal form and several different mutant forms. These different forms of a gene are called **alleles**.

> An **allele** is one of two or more different forms of a gene that codes for a particular polypeptide or protein.

Homozygote and heterozygote
An organism that has two identical alleles for a particular gene is a **homozygote**. An organism that has two different alleles for a particular gene is a **heterozygote**.

Dominant and recessive
We can use letters to represent the different alleles of a gene. For example, we could use **F** to represent the normal cystic fibrosis allele, and **f** to represent a mutant allele.

There are three possible combinations of these alleles in a diploid organism: **FF**, **Ff** or **ff**. These are the possible **genotypes** of the organism.

These different genotypes give rise to different **phenotypes** — the observable characteristics of the organism.

Genotype	Phenotype
FF	normal
Ff	normal
ff	cystic fibrosis

A person with the genotype **Ff** is said to be a **carrier** for cystic fibrosis, because they have the cystic fibrosis allele but do not have the condition.

The **Ff** genotype does not cause cystic fibrosis because the **F** allele is **dominant** and the **f** allele is **recessive**. A dominant allele is one that is expressed (has an effect) in a heterozygous organism. A recessive allele is one that is only expressed when a dominant allele is not present.

● The dominant allele should always be symbolised by a capital letter, and the recessive allele by a small letter. The same letter should be used for both (not **F** and **c**, for example).
● If you are able to choose the symbols that you use in a genetics question, then choose ones where the capital and small letter are different in shape, to avoid confusion (not **C** and **c**, for example).

Monohybrid inheritance

This is the inheritance of a single gene. For example, albinism (the lack of melanin in the skin) is caused by a recessive allele, **a**. Imagine that a man with the genotype **Aa** and a woman with the genotype **AA** have children. In their testes and ovaries, gametes are produced. In the man, half of his sperm will contain the **A** allele and half will contain the **a** allele. All of the woman's eggs will contain the **A** allele.

We can predict the likely genotypes of any children that they have using a **genetic diagram**. Genetic diagrams should always be set out like this:

Parents' genotypes   **Aa**   ×   **AA**

Gametes' genotypes   (A) (a)   (A)

Offspring genotypes and phenotypes

eggs

(A)

sperm
(A) | AA normal
(a) | Aa normal

We can therefore predict that there is an equal chance of any child born to them having the genotype **AA** or **Aa** There is no chance they will have a child with albinism.

If both parents have the genotype **Aa**:

Parents' genotypes   **Aa**   ×   **Aa**

Gametes' genotypes   (A) (a)   (A) (a)

Offspring genotypes and phenotypes

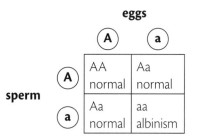

eggs

(A)   (a)

sperm
(A) | AA normal | Aa normal
(a) | Aa normal | aa albinism

We can therefore predict that, each time they have a child, there is a 1 in 4 chance that it will have the genotype **aa** and have albinism.

Notice:

- In the examination, it may be important to show the whole genetic diagram, not just the Punnett square (the table showing the genotypes of the offspring).
- The 'gametes' genotype' line in the genetic diagram shows the *different kinds* of gametes the parents can produce. If there is only one kind of gamete, you only need to write down one kind — there is no need to write the same one down twice. (If you do that, it will not make any difference to your final answer, but will make twice as much work in writing it all down.)
- The gamete genotypes are shown with a circle drawn around them. This is a convention — if you do this, the examiner will understand that they represent gametes.

- There is often a mark for stating the phenotype produced by each genotype among the offspring. The easiest and quickest way to do this is to write the phenotypes in the boxes in the Punnett square.
- The genotypes inside the Punnett square show the chances or *probabilities* of each genotype being produced. If you have four genotypes, as in the example above, this does not mean there will be four offspring. It means that, every time an offspring is produced, there is a 1 in 4 chance it will be **AA**, a 1 in 4 chance it will be **aa** and a 2 in 4 (better written as 1 in 2) chance it will be **Aa**.
- An alternative way of writing 'a 1 in 4 chance' is 'a probability of 0.25', or 'a 25% probability'.
- You could also give the final answer in terms of expected ratios. For example, 'We would expect the ratio of unaffected offspring to offspring with albinism to be 3:1'.
- The chance of any individual child inheriting a particular genotype is unaffected by the genotype of any previous children. Each time a child is conceived the chances are the same, as shown in the genetic diagram above.

> **Expert tip**
>
> It is tempting just to show the Punnett square, but it is almost always important to include the complete genetic diagram in your answer in order to get full marks.

Test crosses

A test cross is a breeding experiment that is carried out to determine the genotype of an organism that shows the dominant characteristic.

For example, in a species of mammal, the gene for hair colour has two alleles, **B** and **b**. Allele **B** gives brown fur and allele **b** gives white fur.

> **Expert tip**
>
> Test crosses used to be called backcrosses, so you may come across this term in old textbooks.

Genotype	Phenotype
BB	brown fur
Bb	brown fur
bb	white fur

If an animal has brown fur, we do not know if its genotype is **BB** or **Bb**. We can find out by breeding it with an animal with white fur, whose genotype must be **bb**.

If there are any white offspring, then the unknown animal must have the genotype **Bb**, as it must have given a **b** allele to these offspring.

If there are no white offspring, then the unknown animal probably has the genotype **BB**. However, it is still possible that it is **Bb** and, just by chance, none of its offspring inherited the **b** allele from it.

Codominance

In the albinism example, one allele is dominant and the other recessive. Some alleles, however, are **codominant**. Each allele has an effect in a heterozygote.

For codominant alleles, it is not correct to use a capital and small letter to represent them. Instead, a capital letter is used to represent the gene, and a superscript to represent the allele.

For example, in some breeds of cattle there are two alleles for coat colour:

C^R is the allele for red coat

C^W is the allele for white coat

> **Expert tip**
>
> It is tempting not to bother writing allele symbols with superscripts because it is much quicker just to use a single letter, but it is very important to do it correctly.

Genotype	Phenotype
$C^R C^R$	red coat
$C^R C^W$	coat with a mixture of red and white hairs — roan
$C^W C^W$	white coat

The inheritance of codominant alleles is shown using a genetic diagram just like the one on p. 143. You might like to try showing that two roan cattle would be expected to have offspring with roan, red and white coats in the ratio 2:1:1.

F1 and F2 generations

When two organisms that are homozygous for two different alleles of a gene, for example **AA** and **aa**, are crossed, the offspring produced are called the F1 generation. The F1 are all heterozygous, i.e. **Aa**.

When two of the F1 generation are crossed, their offspring are called the F2 generation.

Multiple alleles

Many genes have more than two alleles. For example, the gene that determines the antigen on red blood cells, and therefore your blood group, has three alleles — I^A, I^B and I^O.

I^A and I^B are codominant. They are both dominant to I^O, which is recessive.

Genotype	Phenotype
$I^A I^A$	blood group A
$I^A I^B$	blood group AB
$I^A I^O$	blood group A
$I^B I^B$	blood group B
$I^B I^O$	blood group B
$I^O I^O$	blood group O

> **Typical mistake**
>
> F1 cannot be used for the first set of offspring from any cross. You should not use it unless the offspring are the result of a cross between two homozygous parents.

Genetic diagrams involving multiple alleles are constructed in the same way as before. For example, you could be asked to use a genetic diagram to show how parents with blood groups A and B could have a child with blood group O.

> **Expert tip**
>
> Note that you should show the genotypes correctly, using the I symbols with their superscripts. The A and B symbols represent the blood group, not the genotype.

Parents' phenotypes	blood group A	×	blood group B
Parents' genotypes	$I^A I^O$		$I^B I^O$
Gametes' genotypes	(I^A) (I^O)		(I^B) (I^O)

Offspring genotypes and phenotypes

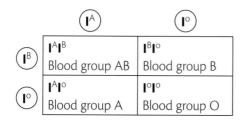

When solving a problem like this:

- Begin by working out and then writing down a table of genotypes and phenotypes, which you can easily refer back to as you work through the problem.
- Consider whether you can tell the genotype of any of the individuals in the problem from their phenotype. Here, we know that the child with blood group O must have the genotype $I^O I^O$.
- Work back from there to determine the genotypes of the parents. In this case, each of them must have had an I^O allele to give to this child.
- Always show a complete genetic diagram. Do not take short cuts, even if you can see the answer straight away, as there will be marks for showing each of the steps in the diagram.

> **Now test yourself**
>
> 2 Construct a genetic diagram to predict the genotypes and phenotypes of the children of two people with blood group AB.
>
> **Answer on p.204**
>
> Tested

Sex linkage

In a human cell, there are two sex chromosomes. A woman has two X chromosomes. A man has an X chromosome and a Y chromosome.

The X chromosome is longer than the Y chromosome. Most of the genes on the X chromosome are not present on the Y chromosome. These are called sex-linked genes.

For example, a gene that determines the production of red-receptive and green-receptive pigments in the retina of the eye is found on the X chromosome. There is a recessive allele of this gene that results in red-green colour-blindness.

A woman has two copies of this gene, because she has two X chromosomes. A man has only one copy, because he has only one X chromosome.

If the normal allele is **A**, and the recessive abnormal allele is **a**, then these are the possible genotypes and phenotypes a person may have:

> A **sex-linked** gene is one that is present on the X chromosome but not on the Y chromosome.

Genotype	Phenotype
$X^A X^A$	female with normal vision
$X^A X^a$	female with normal vision
$X^a X^a$	female with colour blindness
$X^A Y$	male with normal vision
$X^a Y$	male with colour blindness

Notice:
- Sex-linked genes are shown by writing the symbol for the allele as a superscript above the symbol for the X chromosome.
- There are three possible genotypes for a female, but only two possible genotypes for a male.

For example, you could be asked to predict the chance of a woman who is a carrier for colour blindness (that is, heterozygous) and a man with normal vision having a colour-blind child.

Parents' phenotypes	woman with normal vision	✕	man with normal vision

Parents' genotypes	$X^A X^a$		$X^A Y$

Gametes' genotypes (X^A) (X^a) (X^A) (Y)

Offspring genotypes and phenotypes

	(X^A)	(X^a)
(X^A)	$X^A X^A$ Female with normal vision	$X^A X^a$ Female with normal vision
(Y)	$X^A Y$ Male with normal vision	$X^a Y$ Colour-blind male

There is therefore a 1 in 4 chance that a child born to this couple will be a colour-blind boy. There is a 1 in 2 chance of any boy that is born being colour-blind.

Notice:

- Always show the X and Y chromosomes when working with sex-linked genes.
- A boy cannot inherit a sex-linked gene from his father, because he only gets a Y chromosome from his father. His X chromosome (which carries sex-linked genes) comes from his mother.

Another example of sex linkage in humans is haemophilia. This is caused by a recessive allele of a gene for a blood-clotting factor, carried on the X chromosome. In males who have this recessive allele, blood clotting does not take place normally, so wounds can bleed copiously, and internal bleeding can take place, for example into joints.

Now test yourself

3 Explain why a man who is colour-blind cannot pass this condition to his son.

Answer on p.204

Tested

Dihybrid inheritance

Revised

This involves the inheritance of two genes. For example, a breed of dog may have genes for hair colour and leg length.

Allele **A** is dominant and gives brown hair. Allele **a** is recessive and gives black hair.

Allele **L** is dominant and gives long legs. Allele **l** is recessive and gives short legs.

As before, always begin by writing down all the possible genotypes and phenotypes.

Genotype	Phenotype
AALL	brown hair, long legs
AaLL	brown hair, long legs
aaLL	black hair, long legs
AALl	brown hair, long legs
AaLl	brown hair, long legs
aaLl	black hair, long legs
AAll	brown hair, short legs
Aall	brown hair, short legs
aall	black hair, short legs

Notice:

- Always write the two alleles of one gene together. Do not mix up alleles of the two different genes.
- Always write the alleles in the same order. Here, we have decided to write the alleles for hair colour first, followed by the alleles for leg length. Do not swap these round part-way through.

A dihybrid cross

A genetic diagram showing a dihybrid cross is set out exactly as for a monohybrid cross. The only difference is that there will be more different types of gamete. Each gamete will contain just one allele of each gene.

This genetic diagram shows the offspring we would expect from a cross between two dogs that are both heterozygous for both genes.

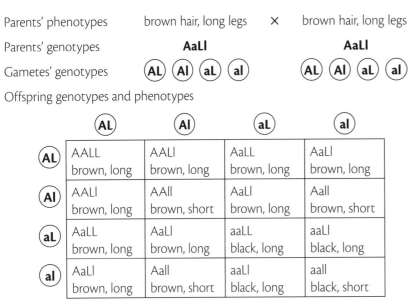

	(AL)	**(Al)**	**(aL)**	**(al)**
(AL)	AALL brown, long	AALl brown, long	AaLL brown, long	AaLl brown, long
(Al)	AALl brown, long	AAll brown, short	AaLl brown, long	Aall brown, short
(aL)	AaLL brown, long	AaLl brown, long	aaLL black, long	aaLl black, long
(al)	AaLl brown, long	Aall brown, short	aaLl black, long	aall black, short

We would therefore expect offspring in the ratio 9 brown hair, long legs : 3 brown hair, short legs : 3 black hair, long legs : 1 black hair, short legs.

Notice:

- Each gamete contains one allele of each gene (one of either **A** or **a**, and one of either **L** or **l**).
- The ratio 9:3:3:1 is typical of a dihybrid cross between two heterozygotes, where the alleles of both genes show dominance (as opposed to codominance).

Autosomal linkage

If two genes are on the same chromosome, then their alleles do not assort independently during meiosis. The alleles on the same chromosome tend to be inherited together. This is called **autosomal linkage** (Figure 16.10).

For example, imagine that in a species of animal, the gene for eye colour (**E/e**) and the gene for fur colour (**F/f**) are on chromosome 3. A male animal has the genotype **EeFf**. In his cells, one of his chromosome 3s carries the **E** and **F** alleles, and the other has the **e** and **f** alleles.

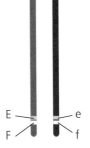

E ――――― e
F ――――― f

Figure 16.10 Autosomal linkage

When meiosis takes place in the male parent's testes, most of the sperm produced will either have the genotype **EF** or **ef**. Only a very small number, produced when crossing over took place between the two gene loci, will have the genotype **Ef** or **eF**.

Now imagine that this male mates with a female with the genotype **eeff**. All of her eggs have the genotype **ef**.

Parents' phenotypes   brown hair, long legs   ×   brown hair, long legs

Parents' genotypes   **AaLl**   **AaLl**

Gametes' genotypes   (AL) (Al) (aL) (al)   (AL) (Al) (aL) (al)

Offspring genotypes and phenotypes

> ## Now test yourself
>
> **4** Use a genetic diagram to show that, when a homozygous recessive individual is crossed with one that is heterozygous for both genes, the expected ratio would be 1:1:1:1.
>
> ## Answer on p.204
>
> Tested

> **Autosomal linkage** is the tendency of two characteristics to be inherited together, because the genes that cause them are on the same (non-sex) chromosome.

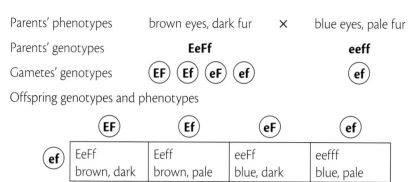

Parents' phenotypes   brown eyes, dark fur   ×   blue eyes, pale fur

Parents' genotypes   **EeFf**   **eeff**

Gametes' genotypes   (EF) (Ef) (eF) (ef)   (ef)

Offspring genotypes and phenotypes

	(EF)	(Ef)	(eF)	(ef)
(ef)	EeFf brown, dark	Eeff brown, pale	eeFf blue, dark	eeff blue, pale

If there was no autosomal linkage, we would expect offspring genotypes in the ratio 1:1:1:1. However, because the genes are linked together, we find that most of the offspring resemble their two parents. Only a small number (perhaps none at all) show different combinations of eye colour and fur colour. These few are called **recombinants**.

Now test yourself

5 Construct a genetic diagram to predict the genotypes and phenotypes of the offspring if the male with genotype **EeFf** is mated with a female with the same genotype.

Answer on p.205

Tested

Using statistics in genetics

 Revised

Genetic diagrams can be used to tell us the probabilities of particular genotypes occurring in offspring. However, because these are just *probabilities*, the actual phenotypic ratios are rarely exactly as we predicted.

For example, imagine that a plant has genes for flower colour (allele **Y** for yellow and **y** for white flowers) and petal size (**P** for large petals and **p** for small petals).

We cross two plants that are heterozygous for both alleles. If these alleles are not linked (i.e. the genes are on different chromosomes) we would expect to get offspring with phenotypes:

yellow, large	yellow, small	white, large	white, small
9 :	3 :	3 :	1

However, what we actually get are 847 offspring with these phenotypes:

yellow, large	yellow, small	white, large	white, small
489 :	151 :	159 :	48

What we would *expect* to have got is:

yellow, large	yellow, small	white, large	white, small
476 :	159 :	159 :	53

The question we need to ask is: are these results close enough to the expected results to indicate that we are right in thinking that the two genes are not linked? Or are they so different from the results we expected that they indicate that something different is happening? In other words, is the difference between our observed and expected results **significant**?

We can use statistics to tell us how likely it is that the difference between our observed results and our expected results is just due to chance, or whether it is so different that we must have been wrong in our predictions.

Null hypothesis and probability level

A statistics test begins by setting up a **null hypothesis**. We then use the statistics test to determine the probability of the null hypothesis being true.

In this case, our null hypothesis would be:

The observed results are not significantly different from the expected results.

We then work through a statistics test, which gives us a *probability of our null hypothesis being correct*. In biology, it is conventional to say that:

Now test yourself

6 Construct two genetic diagrams to show the expected offspring ratios from this cross (a) if the genes are linked with YP on one chromosome and yp on the other, and (b) if the genes are not linked.

Answer on p.205

Tested

A **null hypothesis** is a statement that there is no relationship between two variables, or that the results we have obtained show no significant difference.

- if the probability of the null hypothesis being correct is *greater than or equal to 0.05*, then we can *accept* the null hypothesis.
- if the probability is *less than 0.05* that the null hypothesis is correct, then we must *reject* it

The chi-squared test

To test the possibility of our null hypothesis being correct, the most suitable statistical test for this set of results is the **chi-squared test**. This can also be written as χ^2 **test**.

- Construct a table similar to the one below. You need one column for each category of your results.

	Yellow, large	Yellow, small	White, large	White, small
Observed numbers, O				
Expected numbers, E				
$O - E$				
$(O - E)^2$				
$\dfrac{(O - E)^2}{E}$				
$\dfrac{(O - E)^2}{E}$				

- Fill in the observed numbers and the expected numbers for each category.
- Calculate $O - E$ for each category.
- Calculate $(O - E)^2$ for each category.
- Calculate $\dfrac{(O - E)^2}{E}$ for each category.
- Add together all of the $\dfrac{(O - E)^2}{E}$ values, to give you $\Sigma\dfrac{(O - E)^2}{E}$.

Your table now looks like this:

	Yellow, large	Yellow, small	White, large	White, small
Observed numbers, O	489	151	159	48
Expected numbers, E	476	159	159	53
$O - E$	13	−8	0	−5
$(O - E)^2$	169	64	0	25
$\dfrac{(O - E)^2}{E}$	0.36	0.40	0	0.47
$\Sigma\dfrac{(O - E)^2}{E} = 1.23$				

The number 1.23 is the chi-squared value.

You now need to look this up in a probability table, to find out what it tells us about the probability of the null hypothesis being correct.

This is a part of a probability table for chi-squared values.

Degrees of freedom	Probability of null hypothesis being correct			
	0.1	0.05	0.01	0.001
1	2.71	3.84	6.64	10.83
2	4.60	5.99	9.21	13.82
3	6.25	7.82	11.34	16.27
4	7.78	9.49	13.28	18.46

The numbers inside the cells in the table are chi-squared values.

The numbers in the first column are degrees of freedom. You have to choose the correct row in the table for the number of degrees of freedom in your data. In general:

degrees of freedom = number of different categories − 1

In this case, there were four categories (the four different phenotypes), so there are 3 degrees of freedom.

- Look along the row for 3 degrees of freedom until you find the number closest to the chi-squared value you have calculated. This was 1.23, and all the numbers in the row are much bigger than this. The closest is 6.25.
- Look at the probability associated with this number. It is 0.1. This means that if the chi-squared value was 6.25, there is a 0.1 probability that the null hypothesis is correct.
- Remember that, if the probability of the null hypothesis being correct is equal to or greater than 0.05, you can accept it as being correct. 0.1 is much bigger than 0.05, so you can definitely accept the null hypothesis as being correct. Indeed, because your value of chi-squared was actually much smaller than 6.25, we would need to go a long way further left in the table, which would take us into probabilities even greater than 0.1.
- We can therefore say that the difference between the observed results and expected results is *not significant*. The differences between them are just due to random chance.

Gene interactions

Sometimes, two genes at different loci interact to produce a phenotype. For example, in mice, two genes affect fur colour:

- Gene for distribution of melanin in hairs: **A** produces banding (agouti), **a** produces a uniform distribution.
- Gene for presence of melanin in hairs: **B** produces melanin, **b** does not produce melanin.

Clearly, gene **A/a** can only have an effect if the pigment melanin is present. If there is no melanin, fur colour is white.

Genotype	Phenotype
AABB	agouti
AaBB	agouti
aaBB	black
AABb	agouti
AaBb	agouti
aaBb	black
AAbb	white
Aabb	white
aabb	white

Now test yourself

7 A species of mammal has either white or grey fur, and either a long or short tail. Several crosses between a heterozygous animal with white fur and long tail, and an animal with grey fur and short tail, produced these offspring:

white fur, long tail	12
white fur, short tail	2
grey fur, long tail	2
grey fur, short tail	12

Use the chi-squared test to determine whether these results show that the genes for fur colour and eye colour are on the same chromosome.

Answer on p.205

Tested

Now test yourself

8 Construct a genetic diagram to predict the ratios of phenotypes resulting from a cross between two mice with the genotypes **AaBB** and **AaBb**.

9 How could you use a test cross to find out the genotype of a white mouse?

Answers on p.206

Tested

Gene mutation

Revised

A mutation is a random, unpredictable change in the DNA in a cell. Gene mutation occurs when there is a change in the sequence of bases in one part of a DNA molecule. This can occur by substitution, deletion or insertion (Figure 16.11).

Original DNA

A T G C C G A C T ...

After addition	**After deletion**	**After substitution**
A T T G C C G A C T ...	A T C C G A C T ...	A T C C C G A C T ...

Figure 16.11 Types of gene mutation

Deletion or addition of a base causes all of the following DNA triplets to be different; this is called a frame shift. Frame shifts have a major effect during protein synthesis, because a completely new sequence of amino acids will be used. One of the new triplets could even be a 'stop' code. Substitution, however, may have no effect at all, because only one triplet is altered. This may code for the same amino acid as the original one. Even if it codes for a different one, only that amino acid is changed, and the protein that is made may still be able to function normally.

Mutations are most likely to occur during DNA replication, for example when a 'wrong' base might slot into position in the new strand being built. Almost all these mistakes are repaired immediately by enzymes, but some may persist.

Conditions resulting from mutations in humans include the following:

- **Sickle cell anaemia** — results from a substitution in the gene coding for the β polypeptide in a haemoglobin molecule. You can find more about this on p. 156.
- **Haemophilia** — results from a mutation in the gene coding for a clotting factor, factor VIII. The mutant allele is recessive. This gene is found on the X chromosome, so haemophilia is a sex-linked condition.
- **Albinism** — results from one of several different mutations in a gene coding for melanin production. The mutant allele is recessive, and does not allow melanin to be produced. Mammals homozygous for such an allele have no melanin in their coats or irises, and are white with pink eyes. There are several different mutant alleles that can have this effect. One of them prevents the formation of the enzyme tyrosinase, which is involved in the conversion of the amino acid tyrosine to melanin.

$$\text{tyrosine} \xrightarrow{\text{tyrosinase}} \text{DOPA} \rightarrow \text{dopaquinone} \rightarrow \text{melanin}$$

- **Huntington's disease** — results from a mutation in the gene coding for a protein called huntingtin. The mutant allele is dominant, and affects neurones in the brain. These effects are usually not seen until the person is between 35 and 44 years old, when they begin to develop problems with muscle coordination and cognitive difficulties.

Typical mistake

Students often say that mutation is most likely to happen during meiosis. This is not correct; genes are most likely to mutate during DNA replication. Meiosis does produce variation, but this is by reshuffling genes, not by producing new ones by mutation.

Control of gene expression

All the cells in your body contain a full set of genes. However, in each cell only a certain set of these genes is used to make proteins (expressed). Different genes are expressed in different cells, and at different times.

Gene expression is the transcription and translation of the gene, resulting in the production of the polypeptide or protein for which it codes.

The *lac* operon

Revised

One of the best-studied examples of the control of gene expression is the gene for the production of the enzyme lactase (β galactosidase) in the bacterium *Escherichia coli*. The bacterium only produces lactase when the sugar lactose is present. The gene for this enzyme is closely associated with several other regions of DNA that control whether or not the gene is used to make lactase.

The stretch of DNA that codes for the production of the enzyme is called the *lac* operon (Figure 16.12). It is made up of:

- three **structural genes**, which code for proteins involved in producing the enzyme
- a promoter that allows transcription of the structural genes to begin
- an operator that allows a molecule called a **transcription factor** to bind, which in this case stops the structural genes being transcribed

Part of the bacterium's DNA

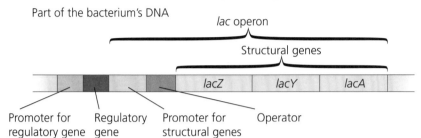

Figure 16.12 The *lac* operon

Alongside the *lac* operon are two other regions of DNA that control the expression of the structural genes:

- a **regulatory gene**, which codes for a protein that acts as a repressor
- a promotor for the regulatory gene, which allows its transcription PIF to begin

When there *is no lactose* in the medium in which the bacterium is growing, the regulatory gene is expressed and therefore the repressor protein is produced. This repressor protein binds to the operator region of the *lac* operon. This prevents RNA polymerase binding to the promoter for the structural genes, so they are not transcribed and no lactase is made.

When there *is lactose* in the medium, it binds to the repressor protein and alters its shape. This prevents the repressor protein from binding to the operator region. Now RNA polymerase can bind to the promoter, and the structural genes can be transcribed. Lactase is made.

Inducible and repressible enzymes

Lactase is said to be an **inducible enzyme**. This means that its production only occurs when its substrate is present. The presence of the substrate prevents the repressor binding with the operator, so the presence of the substrate induces (causes) production of the enzyme. Other enzymes might be **repressible enzymes**. This means that the presence of a particular substance allows the repressor to bind with the operator and stops the enzyme being produced.

Now test yourself

10 Suggest the advantage to the bacterium of producing lactase only when lactose is present.

Answer on p.206

Tested

Transcription factors in eukaryotes

Revised

A transcription factor is a molecule that binds to DNA and causes or prevents the transcription of a gene. We have seen that transcription factors help to control gene expression in bacteria (prokaryotes), and they also do so in eukaryotes.

For example, production of the enzyme amylase in germinating seeds depends on a transcription factor called PIF (phytochrome interacting protein) binding with the promoter region next to the gene that codes for amylase.

- When no gibberellin is present, PIF is bound to a protein called a DELLA protein. This prevents PIF binding with the promoter for the amylase gene, so no amylase is made.
- When gibberellin is present, it binds with a receptor and an enzyme. This activates the enzyme, which breaks down the DELLA protein. PIF is now free to bind to the promoter for the amylase enzyme, and amylase is produced.

This explains how gibberellin causes amylase production during seed germination, described on p. 136.

Revision activity

- Construct a simple diagram showing how gibberellin promotes the synthesis of amylase.

17 Selection and evolution

Variation

Differences between the phenotypes of individuals belonging to the same species are known as variation. Variation can be continuous or discontinuous.

- **Continuous variation** occurs when a feature can have any value between two extremes, such as height and mass in humans.
- **Discontinuous variation** occurs when a feature has only a few distinct categories, such as blood groups in humans.

In general, continuous variation is caused when the effects of many different genes act together to influence a characteristic. The environment may also have an effect. Discontinuous variation is caused by one or a few genes that control the characteristic, with little or no influence from the environment.

Effect of the environment on phenotype
Revised

Hair colour in cats

Many different genes determine hair colour in cats. At least eight different genes, at different loci, are known to influence hair colour and it is thought that there are probably more. These are known as **polygenes**. Depending on the particular combination of alleles that a cat has for each of these genes, it can have any of a very wide range of colours. Hair colour in cats is an example of **continuous variation**. There is a continuous range of variation in colour between the very lightest and very darkest extremes.

The cat hair colour genes exert their effect by coding for the production of enzymes. One such gene is found at the **C** locus. Siamese cats have two copies of a recessive allele of this gene called c^s. This gene codes for an enzyme that is sensitive to temperature. It produces dark hair at the extremities of paws, ears and tail where the temperature is lower, and light hair in warmer parts of the body. The colouring of a Siamese cat is therefore the result of interaction between genes and environment.

Height

Human height is also affected by many different genes at different loci. It is also strongly affected by environment. Even if a person inherits alleles of these genes that give the potential to grow tall, he or she will not grow tall unless the diet supplies plenty of nutrients to allow this to happen. Poor nutrition, especially in childhood, reduces the maximum height that is attained. A similar situation occurs in plants, where individual plants with genes that allow tall growth may remain stunted if they are growing in soil that is short of particular mineral ions (e.g. nitrate) or if they do not get enough water.

Cancer

The risk of developing cancer is influenced by both genes and environment. For example, a woman with particular alleles of the genes *BRCA1* or *BRCA2* has a 50–80% of chance of developing breast cancer at some stage in her life. This is a much higher risk than for people who do not have these alleles. The normal alleles of these genes protect cells from changes that could lead to them becoming cancerous.

However, environment also affects this risk. Smoking, for example, increases the risk even further. Taking the drug tamoxifen can reduce the risk.

The *t*-test
Revised

You may be asked to use the *t*-test to determine whether the variation shown in one population is significantly different from the variation in another population. The *t*-test is described on pp. 186–188.

Natural and artificial selection

In a population of organisms, each parent can produce more young than are required to maintain the population at a constant level. For example, each pair of adult foxes in a population needs to have only two young in their lifetime to keep the population the same size. However, a pair of foxes can have more than 30 young. The population can **potentially overproduce**. But the population generally stays approximately the same size because most of the young do not survive long enough to reproduce.

In a population, not every organism has exactly the same alleles or exactly the same features. There will be **genetic variation** within the population. Those organisms whose particular set of features is best suited to the environment are most likely to survive. Those with less useful features are more likely to die.

The organisms with the most useful features are therefore more likely to reach adulthood and reproduce. Their alleles will be passed on to their offspring. Over many generations, the alleles that confer useful characteristics on an individual are therefore likely to become more common in the population. Alleles that do not produce such useful characteristics are less likely to be passed on to successive generations and will become less common.

This process is called **natural selection**. Over time, it ensures that the individuals in a population have features that enable them to survive and reproduce in their environment.

> **Natural selection** is the increased chance of survival and reproduction of organisms with particular phenotypes, because they are better adapted to their environment than those with other phenotypes.

Stabilising, directional and disruptive selection
Revised

When the environment is fairly stable, natural selection is unlikely to bring about change. If the organisms in a population are already well adapted to their environment, then the most common alleles in the population will be those that confer an advantage on the organisms, and it is these alleles that will continue to be passed on to successive generations. This is called **stabilising selection**.

However, if the environment changes, alleles that were previously advantageous may become disadvantageous. For example, in a snowy environment the individuals in a species of mammal may have white fur that camouflages them against the snow and confers an advantage in escaping predators. If the climate changes so that snow no longer lies on the ground, then animals with white fur may be more likely to be killed than animals with brown fur. Those with brown fur are now most likely to reproduce and pass on their alleles to the next generation. Over time, brown may become the most common fur colour in the population. This is an example of **directional selection** or **evolutionary selection** (Figure 17.1).

Directional selection may result in **evolution**. Evolution can be defined as a long-term change in the characteristics of a species, or in the frequency of particular alleles within the species.

In a snowy environment, white animals are more camouflaged, making them less visible to predators, so are at a selective advantage. The white animal is most likely to survive.

In an environment without snow, white animals are very noticeable and more likely to be preyed upon. The white animal is least likely to survive.

Figure 17.1 Stabilising and directional selection

In some circumstances, two different phenotypes can each confer advantages. For example, in a population of snails, those with dark brown shells and those with pale shells may each be well camouflaged, while those with mid brown shells may not be camouflaged. In this case, selection will tend to remove most of the snails with mid-range colour from the population, resulting in two distinct colour ranges surviving. This is called **disruptive selection**.

> **Now test yourself**
>
> 1 Sketch a graph, like those showing the results of stabilising and directional selection, to show the result of disruptive selection.
>
> **Answer on p.206**
>
> Tested

The founder effect and genetic drift

Revised

Natural selection is not the only way in which allele frequencies can change in a population. For example, imagine that just three lizards were washed away from an island and drifted to another island where there were no lizards. These three lizards would not contain all of the various alleles present in the original population. The population that grew from these three colonisers would therefore have different allele frequencies from the original population. This is called the **founder effect**.

In small populations, some alleles may be lost just by chance, because the few individuals that have them just happen not to reproduce. This can also happen to relatively rare alleles in larger populations. This is called **genetic drift**.

Sickle cell anaemia and malaria

Revised

Similarities between the global distribution of the genetic disease sickle cell anaemia, and the infectious disease malaria, are a result of natural selection. This illustrates the effect of environmental factors on allele frequencies.

- Sickle cell anaemia is caused by an allele of a gene coding for the β polypeptide in haemoglobin. This is described on p. 47.
- Malaria is caused by a protoctist, *Plasmodium*, which is transmitted by *Anopheles* mosquitoes. This is described on pp. 64–65.

A person who is homozygous for the faulty allele for the β polypeptide, **Hb^{S}Hb^S**, has sickle cell anaemia. This is a serious disease and a child with this genotype is unlikely to survive to adulthood and have children.

A person who is homozygous for the normal allele for the β polypeptide, **Hb^{A}Hb^A**, is vulnerable to malaria. In parts of the world where the *Anopheles* mosquitoes that transmit malaria are found, and where malaria is common, a child with this genotype may not survive to adulthood and have children.

A person who is heterozygous for these alleles, **Hb^{A}Hb^S**, does not have sickle cell anaemia. They are also much less likely to suffer from a serious attack of malaria than a person with the genotype **Hb^{A}Hb^A**. Therefore a heterozygous person is the most likely to survive to adulthood and have children.

The allele **Hb^S** therefore continues to survive in populations where malaria is present, because it is passed on from generation to generation by heterozygous people. In parts of the world where malaria is not common, there is no selective advantage in having this allele, and it is very rare (Figure 17.2).

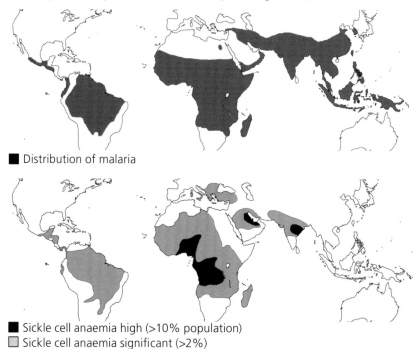

■ Distribution of malaria

■ Sickle cell anaemia high (>10% population)
□ Sickle cell anaemia significant (>2%)

Figure 17.2 The global distribution of malaria and sickle cell anaemia

The Hardy–Weinberg principle

Revised

The Hardy–Weinberg principle states that, as long as certain conditions are fulfilled, the frequency of a particular allele is likely to remain approximately the same over many generations. These conditions are as follows:

- The population is large.
- There is random mating between individuals in the population.
- There are no significant selection pressures that give particular genotypes an advantage or disadvantage.
- There are no new mutations.
- There is no introduction of new alleles by immigration.

If all of these conditions are fulfilled, we can use the Hardy–Weinberg equations to calculate the frequencies of two alleles of a gene, for example **A** and **a**, in a population.

Equation 1

$$p + q = 1$$

where p is the frequency of allele **A** in the population and q is the frequency of allele **a**.

Frequencies are expressed as decimals. For example, if there are equal numbers of each allele in the population, then p and q are both 0.5.

Equation 2

$$p^2 + q^2 + 2pq = 1$$

This equation represents the frequencies of the different genotypes in the population:

- p^2 is the frequency of individuals with genotype **AA**
- q^2 is the frequency of individuals with genotype **aa**
- $2pq$ is the frequency of individuals with genotype **Aa**

Using the Hardy–Weinberg equations

These equations can be used to calculate the frequencies of the alleles in a population. For example, let us say that 2 in every 100 people in a population have a genetic conditions that we know is caused by a recessive allele, **a**. This frequency can be written as 0.02. These people must all be homozygous for this recessive allele, so they are represented by q^2. So:

$q^2 = 0.02$

So:

$q = \sqrt{0.02} = 0.14$

$p + q = 1$

So:

$p = 1 - 0.14 = 0.86$

We can therefore say that the frequency of the recessive allele **a** in the population is 0.14 (which is the same as 14%) and the frequency of the dominant allele **A** in the population is 0.86 (86%).

We can also calculate the frequency of carriers in the population. Their frequency is $2pq$, which is:

$2 \times 0.14 \times 0.86 = 0.24$

We would therefore expect just under one quarter of individuals in the population to be carriers of the recessive allele **a**.

Now test yourself

2   In a population of 5000 animals, 100 have a genetic condition caused by a recessive allele. Calculate:

   a the frequency of the recessive allele in the population

   b the frequency of heterozygous animals in the population

Answers on p.206

Tested

Selective breeding

Revised

Selective breeding is the choice, by humans, of which animals or plants are allowed to breed. This process is sometimes known as **artificial selection**.

Breeding cattle with high milk yields

Dairy cattle are breeds of cattle that are kept for milk production. Only females (cows) produce milk. Selective breeding is therefore done by using cows that have high milk yields and bulls whose female relatives (mothers, sisters) have high milk yields.

If a particular bull is proved to have many female offspring with high milk yields, then semen can be taken from him and used to inseminate many cows, using artificial insemination procedures.

Selective breeding has produced massive increases in the volume and quality of milk produced by cows (Figure 17.3).

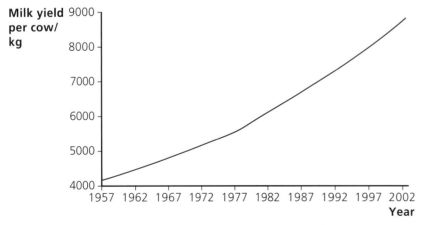

Figure 17.3 Mean milk yield per cow in Sweden between 1957 and 2002

However, such intensive selective breeding can inadvertently cause health problems for the cattle. For example, producing so much milk can weaken their general health. Carrying such large quantities of milk in their udders (mammary glands) can cause problems with leg joints.

Breeding disease-resistant varieties of wheat and rice

Farmers may want to have wheat or rice plants that produce large yields of grain and that are resistant to fungal disease such as rust. They want the plants to be short, so that more energy can be put into growing grain (seed) rather than wasted on stems, and also so that the plants are less likely to fall over in heavy rain or high wind.

Breeders can choose a wheat or rice plant that is short, with high yields of grain, and another that does not have these characteristics but is very resistant to rust. They take pollen from one plant and place it on the stigmas of the other. (The anthers of the second plant are first removed, so that it cannot pollinate itself.)

The resulting seeds are collected and sown. The young plants are allowed to grow to adulthood. Then the plants that show the best combination of desired characteristics are bred together. This continues for several generations, until the breeder has a population of plants that have high yield and high resistance to rust.

Breeding dwarf varieties of crop plants

We have seen that gibberellin causes plant stems to grow longer. Some varieties of crop plants have mutant alleles of the gibberellin gene, and do not produce as much gibberellin. These dwarf plants are often higher yielding than tall varieties, because they do not 'waste' energy in growing tall. They may also be more able to stand upright in strong winds or heavy rain. Selective breeding can be used to combine this characteristic with high yield.

Inbreeding and hybridisation in maize

Farmers require crop plant varieties in which all the individuals are genetically identical. This means that, if the seed is sown at the same time and in the same conditions, then the plants will all grow uniformly. They will ripen at the same time and grow to the same height, which makes harvesting easier. The grain will all have similar characteristics, which makes marketing easier.

Genetic uniformity is usually achieved through **inbreeding** (breeding a plant with itself, or with other plants with the same genotype) for many generations. However, in maize, inbreeding results in weak plants with low yields. This is called **inbreeding depression**.

Maize breeding therefore involves producing **hybrids** between two inbred lines. Like most selective breeding of cereal crops, it is done by commercial organisations, not by farmers themselves.

- Inbred lines (genetically uniform populations) of maize with desirable characteristics are identified, and crossed with other inbred lines with different sets of desirable characteristics.
- The resulting hybrids are assessed for features such as yield, ability to grow in dry conditions, and resistance to insect pests. The best of these hybrids are then chosen for commercial production.
- Large quantities of the two inbred lines from which the hybrid was bred are grown, as it is from these that all the seed to be sold will be produced.
- Each year, the two inbred lines are bred together, and the seed collected from them to sell as hybrid seed.

This method of breeding avoids problems of inbreeding depression. The hybrid plants are genetically uniform (although they will be heterozygous for several genes) because they all have the same two inbred parents.

Notice:

- Artificial selection is similar to natural selection, in that individuals with particular characteristics are more likely to breed than others.
- However, in artificial selection, individuals without these characteristics will not breed at all, whereas in natural selection there is still a chance that they might breed.
- Artificial selection may therefore produce bigger changes in fewer generations than natural selection normally does.
- The breeders do not need to know anything about the genes or alleles that confer the characteristics they want their plants to have; they just choose the plants with those characteristics and hope that the appropriate alleles will be passed on to the next generation.
- Artificial selection usually requires many generations of selection before the desired result is obtained.

Evolution

The theory of evolution states that species have changed over time. One species can, over time, give rise to one or more new species.

> **Evolution** is the change in the characteristics of a population or a species over time.

Molecular evidence for evolutionary relationships

There is a large quantity of evidence to support the theory of evolution. One example of this evidence is the comparison of base sequences in DNA, or amino acid sequences in proteins, in different organisms.

Mitochondria contain a single DNA molecule that is passed on down the female line. Analysis of mitochondrial DNA (mtDNA) can be used to determine how closely related two different species are. The more similar the sequences of bases in their DNA, the more closely related they are considered to be.

Amino acid sequences in proteins can be used in a similar way. For example, the protein cytochrome c, involved in the electron transport chain, is found in a very wide range of different organisms, suggesting that that they all evolved from a common ancestor. Differences in the amino acid sequences in cytochrome c suggest how closely or distantly related particular species are.

Speciation

A **species** is often defined as a group of organisms with similar morphological and physiological characteristics, which are able to breed with each other to produce fertile offspring. So, for example, lions and tigers are distinct species, even though in a zoo they may be persuaded to breed together. Such interbreeding between the two species never occurs in the wild and, in any case, the offspring are not able to breed themselves.

> **Speciation** is the production of one or more new species from an existing species.

So how are new species produced? We have seen that natural selection can produce changes in allele frequency in a species, but how much change is needed before we can say that a new species has been formed?

The crucial event that must occur is that one population must become unable to interbreed with another. They must become **reproductively isolated** from one another. Once this has happened, we can say that the two populations are now different species.

Allopatric speciation

A group of individuals in the population may become **geographically separated** from the rest. If speciation occurs as a result of geographical separation, it is known as **allopatric speciation**.

For example, a few lizards might get carried out to sea on a floating log, and be carried to an island where that species of lizard was not previously found. The lizards on this island are subjected to different environmental conditions from the rest of the species left behind on the mainland. Different alleles are therefore selected for in the two groups. Over time, the allele frequency in the lizards on the island becomes very different from the allele frequency in the original, mainland lizards. This may cause their characteristics to become so different that — even if a bridge appears between the island and the mainland — they can no longer interbreed to produce fertile offspring.

Sympatric speciation

Two groups of individuals living in the same area may become unable to breed together — for example, because one group develops courtship behaviours that no longer match the courtship behaviour of the other group, or because they live in different habitats in the same area (ecological separation). This is known as **sympatric speciation**.

Isolating mechanisms

Possible reasons for the failure of two groups to interbreed include the following:

- They have evolved different courtship behaviours, so that mating no longer occurs between them.
- The sperm of one group are no longer able to survive in the bodies of the females of the other group, so fertilisation does not occur.
- The number or structure of the chromosomes is different, so that the zygote that is formed by fertilisation does not have a complete double set of genes and cannot develop.
- Even if a zygote is successfully produced, the resulting offspring may not be able to form gametes, because its two sets of chromosomes (one from each parent) are unable to pair up with each other successfully and so cannot complete meiosis.

These isolating mechanisms can be divided into pre-zygotic and post-zygotic mechanisms. Pre-zygotic mechanisms act *before* a zygote is formed. Post-zygotic mechanisms act *after* a zygote is formed.

Extinction

The fossil record shows us that many species of organism used to live on Earth but now no longer exist. There are been several times when very large numbers of species became extinct, and these are known as **mass extinctions**. For example, all dinosaur species disappeared from the Earth about 66 million years ago. There is still uncertainty about exactly why this happened, but it is thought to have resulted from severe climate change following a collision of a large asteroid with Earth and/or very extensive volcanic eruptions.

In general, extinctions tend to be caused by:

- climate change; for example, global warming can result in some species not being able to find habitats to which they are adapted for survival
- competition; for example, a newly evolved species, or one that migrates into a new area, may out-compete a resident species, which then becomes extinct
- habitat loss; for example, if humans cut down large areas of rainforest
- direct killing by humans

Now test yourself

3 Which of the listed isolating mechanisms are pre-zygotic, and which are post-zygotic?

Answers on p.206

Tested

18 Biodiversity, classification and conservation

Biodiversity

Biodiversity includes:

- the range of ecosystems or habitats
- the numbers of different species
- the genetic variation that exists within the populations of each species in a particular area

You need to know definitions of species, ecosystem and niche.

> **Biodiversity** is the variety of ecosystems, habitats, species and genotypes that exists in an area.
>
> **Species** (see p. 160).
>
> **Ecosystem** is an interacting system of all the living and non-living things in a defined area.
>
> **Niche** is the role of a species in an ecosystem.

Determining the biodiversity of an area Revised

We can find out something about the biodiversity of an area by measuring the **species richness**. This is the number of different species in the area. The greater the species richness, the greater the biodiversity.

Assessments of biodiversity often involve measuring the **distribution** and **abundance** of the different species. You cannot usually count every single organism in the area, so a **sampling** technique is used. It is important that this is random, to avoid bias in your choice of area in which to make your measurements. For example, you could use a mobile phone app to generate random numbers. These numbers can be used as *x* and *y* coordinates to tell you where to place a **quadrat** within a defined area.

> **Typical mistake**
>
> The correct spelling is quadrat, not quadrant.

Methods of sampling

Frame quadrats

This is a square frame that is placed onto the ground to define an area within which organisms are counted. Within the quadrat, you can either count the actual numbers of organisms (e.g. limpets on a rocky sea shore) or estimate the percentage cover of each species (e.g. different species of plants in meadow).

Line transects

Frame quadrats are often placed randomly (see above) in the area to be studied. However, sometimes you want to know how species abundance changes along a particular line — for example, from the bottom of a beach to the top, or from a heavily shaded area to a sunnier part of a meadow. To do this, you can place quadrats along a **transect**. Lie a long measuring tape on the ground, and then place your quadrats at equal intervals (say, every 2 metres) along the tape.

Belt transects

You can also sample the species at every point along the tape. This can be done by placing your quadrats in succession along the tape, with no gaps between them; or by recording every organism that is touching the tape itself.

Mark–release–recapture

This is a method for estimating the population of species of mobile animal.

- First, a large number of animals is captured. Each one is marked in a way that will not affect its chances of survival. For example, a mouse could have small patch of fur cut away; a woodlouse could have a small spot of paint applied. The number of marked animals is recorded.

- The marked animals are released, and time is allowed for them to spread naturally back among the population.
- A second sample is captured. The number of animals is recorded, and also the number of these that were marked.

The total size of the population can then be estimated by this calculation:

$$\text{number in population} = \frac{\text{number in first sample} \times \text{number in second sample}}{\text{number of marked animals in second sample}}$$

Now test yourself

1 A student caught 38 woodlice and marked them. He released them back into the place from where they were caught. Two days later, he caught 40 woodlice in the same area. Six of them were marked. Calculate the estimated size of the woodlouse population.

Answers on p.206

Tested

Analysing your results

Revised

Spearman's rank correlation

This is a statistical test that can be used to see if there is a correlation between the distribution and abundance of two species, or if there is a correlation between the distribution and abundance of a species, and a particular environmental factor. This test is described on pp. 188–189.

Pearson's linear correlation

This test tells you if there is a linear correlation between the distribution and abundance of two species, or if there is a linear correlation between the distribution and abundance of a species, and a particular environmental factor. This test is described on p. 190.

Simpson's index of diversity (D)

This is calculated using the formula:

$$D = 1 - \Sigma \left(\frac{n}{N}\right)^2$$

where:

D = Simpson's index of diversity

n = the total number of individuals of one species

N = the total number of all individuals of all species in your sample

The greater the value of D, the greater the biodiversity.

Now test yourself

Tested

2 Use Simpson's index of diversity to determine the species diversity of an area from which these results were obtained.

Species	Number of individuals, n
A	2
B	35
C	1
D	81
E	63
F	2
G	5
H	11
I	1
	total number of individuals = 201

Answer on p.206

Classification

Biologists classify organisms according to how closely they believe they are related to one another. Each species has evolved from a previously existing species. We do not usually have any information about these ancestral species, so we judge the degree of relatedness between two organisms by looking carefully at their physiology, anatomy and biochemistry. The greater the similarities, the more closely they are thought to be related.

The system used for classification is a **taxonomic system**. This involves placing organisms in a series of taxonomic units that form a hierarchy. The largest unit is the **domain**. There are three domains: **Archaea**, **Bacteria** and **Eukarya**. The Archaea and Bacteria are prokaryotic (see p. 12 for the features of prokaryotic cells), but are now known to differ greatly in their metabolism. The Eukarya are eukaryotic organisms, many of which are multicellular.

The taxonomic hierarchy

The Eukarya are divided into four **kingdoms**.

Kingdom Protoctista These organisms have eukaryotic cells. They mostly exist as single cells, but some are made of groups of similar cells.

Kingdom Fungi Fungi have eukaryotic cells surrounded by a cell wall, but this is not made of cellulose and fungi never have chloroplasts.

Kingdom Plantae These are the plants. They have eukaryotic cells surrounded by cellulose cell walls and they feed by photosynthesis.

Kingdom Animalia These are the animals. They have eukaryotic cells with no cell wall.

Kingdoms are subdivided into **phyla**, **classes**, **orders**, **families**, **genera** and **species** (Figure 18.1).

Kingdom	Animalia
Phylum	Chordata
Class	Mammalia
Order	Rodentia
Family	Muridae
Genus	*Mus*
Species	*Mus musculus* (house mouse)

Figure 18.1 An example of classification

Note that viruses are not included in the three-domain classification. This is because they are on the borderline between living and non-living things. They have none of the characteristics of living things, except for the ability to reproduce. However, even this can only be done when they have entered a host cell, where they use the host cells' ability to copy, transcribe and translate base sequences, in order to synthesis proteins and nucleic acids to make new viruses. See p. 13 for a diagram of a virus.

Viruses are classified according to the type of nucleic acid that they contain — that is, whether it is DNA or RNA. They can be further classified according to whether the nucleic acid is single-stranded or double-stranded.

Conservation

Conservation aims to maintain biodiversity.

> **Conservation** is the maintenance or increase of biodiversity in an area.

Reasons for maintaining biodiversity
Revised

- **Stability of ecosystems** The loss of one of more species within a community may have negative effects on others, so that eventually an entire ecosystem becomes seriously depleted.
- **Benefits to humans** The loss of a species could prevent an ecosystem from being able to supply humans with their needs.
  - For example, an ancient civilisation in Peru, the Nazca people, are thought to have cut down so many huarango trees that their environment became too dry and barren for them to grow crops.
  - Today we make use of many different plants to supply us with drugs. There may be many more plant species, for example in rainforests, that could potentially provide life-saving medicines.
  - In some countries, such as Kenya, people derive income from tourists who visit the country to see wildlife.
  - Several scientific studies have found that people are happier and healthier if they live in an environment with high biodiversity.
- **Moral and ethical reasons** Many people feel that it is clearly wrong to cause extinction of a species, and that we have a responsibility to maintain environments in which all the different species on Earth can live.

Threats to biodiversity
Revised

We have seen (p. 161) that there have been mass extinction events in the past, when huge numbers of species became extinct over a relatively short period of time. We are currently experiencing another potential mass extinction event. Much of this is due to human activity. For example:

- Increases in greenhouse gases (mostly carbon dioxide and methane) are causing global warming. Much of this increase is caused by human activities, such as burning fossil fuels. Global warming is changing the environmental conditions in many habitats, including raised temperatures, changes in rainfall patterns and a higher frequency of extreme events such as hurricanes. Some species may no longer be able to find suitable habitats in which to live, and will become extinct.
- As the human population continues to increase, we make ever-increasing demands on the environment for resources. Land that could support natural ecosystems and high biodiversity is taken over by humans for farming, building towns, mining and building roads.
- Pollutants — for example, heavy metals, acidic gases (such as SO_2, NOx) and phosphates (from detergents) — released into the air, water and soil may change the environment in ways that reduce the chance of survival of organisms, and could make species extinct.
- Introduction of alien species can threaten native species. For example, the introduction of possums into New Zealand has threatened many native birds with extinction. Possums eat large quantities of leaves from native trees, leaving less food for native species and less diversity of trees in forests.

Also see below for the particular reasons for the threat to tigers.

Endangered species

A species is said to be endangered if its numbers have fallen so low that it may not be able to maintain its population for much longer. For example, tigers are seriously endangered. This has happened for many reasons:

● Tigers are hunted by humans for various body parts that are thought to be effective medicines by people in countries such as China and Korea.
● Tigers require large areas of land to live and hunt successfully. Expanding human populations also require land, and this has greatly reduced the area of suitable habitat for tigers.
● Tigers are large predators that can pose threats to humans and their livestock, and so may be killed.

It is generally considered desirable for a species to have reasonably large genetic diversity. This means that if the environment changes — for example, because of climate change or if a new pathogen emerges — then at least some of the population may possess features that will enable them to survive. Genetic diversity allows species to become adapted to a changing environment.

The tiger populations in many areas are now so low that genetic diversity is small. This makes it more likely that an individual tiger will inherit the same recessive allele from both parents, which may produce harmful characteristics. It also makes it less likely that the population will be able to adapt (through natural selection acting on natural variation) to changes in its environment.

Methods of protecting endangered species

Habitat conservation

The best way to conserve threatened species is in their own natural habitat. For example, national parks, marine parks and nature reserves can set aside areas of land or sea in which species are protected. A species will only survive if the features of its habitat that are essential to its way of life are maintained. Maintaining habitat is also likely to be beneficial to many other species in the community.

However, this is often difficult because people living in that area have their own needs. In parts of the world where people are already struggling to survive, it is difficult to impose restrictions on the way they can use the land where they live. The best conservation programmes involve the local people in habitat conservation and reward them for it in some way. For example, they could be given employment in a national park, or could be paid for keeping a forest in good condition.

A habitat that has been severely damaged is said to be **degraded**. It is sometimes possible to restore degraded habitats. For example, a heavily polluted river, or one that has been over-engineered (i.e. its channel has been straightened, or its banks have been concreted over) can be restored to a more natural state.

> **Expert tip**
>
> Try to find out about a local example of a degraded habitat and how it is being restored, so that you could give details of this in an answer to an examination question.

Zoos

Captive breeding programmes This involves collecting together a small group of organisms of a threatened species and encouraging them to breed together. In this way, extinction can be prevented. The breeding programme will try to maintain or even increase genetic diversity in the population by breeding unrelated animals together. This can be done by moving males from one zoo to another, or by using in vitro fertilisation (see below) with frozen sperm transported from males in another zoo. It may also be possible to implant embryos into a surrogate mother of a different species, so that many young can be produced even if there is only a small number of females of the endangered species. Eggs and embryos, like sperm, can be frozen and stored for long periods of time. Collections of frozen sperm, eggs or embryos are sometimes known as 'frozen zoos'.

Reintroduction programmes The best captive breeding programmes work towards reintroducing individuals to their original habitat, if this can be made safe for them. It is very important that work is done on the ground to prepare the habitat for the eventual reintroduction of the animals. For example, the scimitar-horned oryx has been successfully reintroduced to Tunisia, following a widespread captive breeding programme in European zoos and the preparation and protection of suitable habitat, including the education and involvement of people living in or around the proposed reintroduction area.

Education Zoos can bring conservation issues to the attention of large numbers of people, who may decide to contribute financially towards conservation efforts or to campaign for them. Entrance fees and donations can be used to fund conservation programmes both in the zoo itself and in natural habitats.

Research Animals in zoos can be studied to find out more about their needs in terms of food, breeding places and so on. This can help to inform people working on conservation in natural habitats.

Botanic gardens

Botanic gardens are similar to zoos, but for plants rather than animals. Like zoos, they can be safe havens for threatened plant species, and are involved in breeding programmes, reintroduction programmes, education and research.

Seed banks

Seed banks store seeds collected from plants. Many seeds will live for a very long time in dry conditions, but others need more specialised storage environments. A few of the seeds are germinated every so often so that fresh seed can be collected and stored.

Seed banks can help conservation of plants just as zoos can help conservation of animals. The Royal Botanic Gardens at Kew in the UK has a huge seed bank at Wakehurst Place, Sussex, UK. Collectors search for seeds, especially those of rare or threatened species, and bring them to the seed bank where they are carefully stored. Another seed bank, built into the permafrost (permanently frozen ground) in Norway, aims to preserve seeds from all the world's food crops.

Methods of assisted reproduction

It is often desirable to speed up the natural reproduction rate of animals in zoos, in order to increase the number of individuals of the species more rapidly.

IVF

This stands for in vitro fertilisation. 'In vitro' means 'in glass'. This method may be necessary if animals do not reproduce in captivity, perhaps because the conditions are not entirely suitable to stimulate them to do so. It can also enable genetic diversity to be maintained, by transferring frozen sperm from an animal in one zoo, to be used to fertilise eggs from an unrelated female in another zoo.

Eggs are taken from a female. She may be given hormones before this is done, to make her produce more eggs than usual. Sperm is taken from a male. The eggs and sperm are placed in a sterile container, in a suitable liquid, where the sperm will fuse with the eggs and fertilise them. The zygotes that are formed are then inserted into the uterus of the female.

Embryo transfer and surrogacy

IVF may produce more zygotes than can be implanted in one female. Embryos formed from one female's eggs can be inserted into the uteruses of other females, which are known as surrogate mothers. In some cases, the embryos may be put into the uteruses of other, closely related, non-endangered species.

> **Typical mistake**
>
> IVF is often confused with artificial insemination, AI. AI involves artificially introducing sperm (in semen) to a female's reproductive system.

Preventing overpopulation   Revised

In many seemingly natural ecosystems, human activities have removed top predators, such as wolves. This can allow the populations of their prey species, such as deer, to become very numerous. This in turn may threaten other organisms in the ecosystem; for example, herbivores can overgraze the available vegetation, which can result in soil erosion and/or the extinction of plant species.

Conservation may therefore require human intervention to control the populations of a species, even when this species is protected. This can be done by:

- culling, where selected animals in a population are killed. This may be done by hunters, who are given licences allowing them to kill a certain number of animals of a particular sex — for example, deer culling in the USA. Sometimes, culling is done in zoos, where a particular animal is deemed not to be suitable for breeding.
- contraception. For example, in South Africa, female elephants may be given vaccines that cause them to produce antibodies that prevent sperm from fertilising their eggs.

Roles of NGOs   Revised

NGO stands for non-governmental organisation. These are organisations that are not funded by a country's government, but by charitable donations. There are many international NGOs, and most countries also have more local NGOs working actively in conservation. One of the most important international conservation NGOs is WWF, the Worldwide Fund for Nature. It receives donations from individuals and organisations all over the world, and uses them in programmes that include attempts to conserve individual species (for example, snow leopards), to train local people to become wildlife wardens, or to plant native trees to re-establish forests.

International treaties   Revised

Although some conservation can be managed locally, some issues require international cooperation. For example, governments of many countries have signed up to an agreement called CITES, the Convention on International Trade in Endangered Species. This treaty sets out rules that prevent listed species, or products derived from them (such as rhino horn), being moved from one country to another.

19 Genetic technology

Genetic technology, or genetic engineering, is the manipulation of genes in living organisms. Genes extracted from one organism can be inserted into another. This can be done within the same species (for example, in gene therapy) or genes may be transferred from one species to another. Genes can also be synthesised in the laboratory by linking DNA nucleotides together in a particular order, and then inserting this gene into an organism. The transferred gene is then expressed in the recipient organism.

These processes produce recombinant DNA — that is, DNA whose base sequences would not normally be found in an organism.

> **Recombinant DNA** is DNA with base sequences that would not normally be present in an organism; this is usually the result of the introduction of DNA from a different species.

Now test yourself

1 How does gene technology differ from selective breeding?

Answers on p.206

Tested

Tools for genetic technology

PCR

Revised

PCR stands for **polymerase chain reaction**. It is an automated method of making multiple, identical copies (cloning) of a tiny sample of DNA.

PCR involves the exposure of the DNA to a repeating sequence of different temperatures, allowing different enzymes to work. The process shown in Figure 19.1 is repeated over and over again, eventually making a very large number of identical copies of the original DNA molecule.

The **primer** is a short length of DNA with a base sequence complementary to the start of the DNA strand to be copied. This is needed to make the DNA polymerase begin to link nucleotides together as it makes a copy of the exposed DNA strand.

The DNA polymerase that is used in PCR is called *Taq* polymerase. This enzyme was sourced from a bacterium called *Thermophilus aquaticus*, which lives in hot springs in Yellowstone Park in the USA. Its enzymes are stable even at high temperatures, which is why it is suitable for catalysing DNA synthesis in PCR.

Gel electrophoresis

Revised

Electrophoresis is a way of separating:
- strands of DNA of different lengths, or
- polypeptides with different base sequences

It does this by applying an electrical potential difference across a gel in which a sample is placed. Lengths of DNA or polypeptides of different masses, or with different charges, will move across the gel at different speeds.

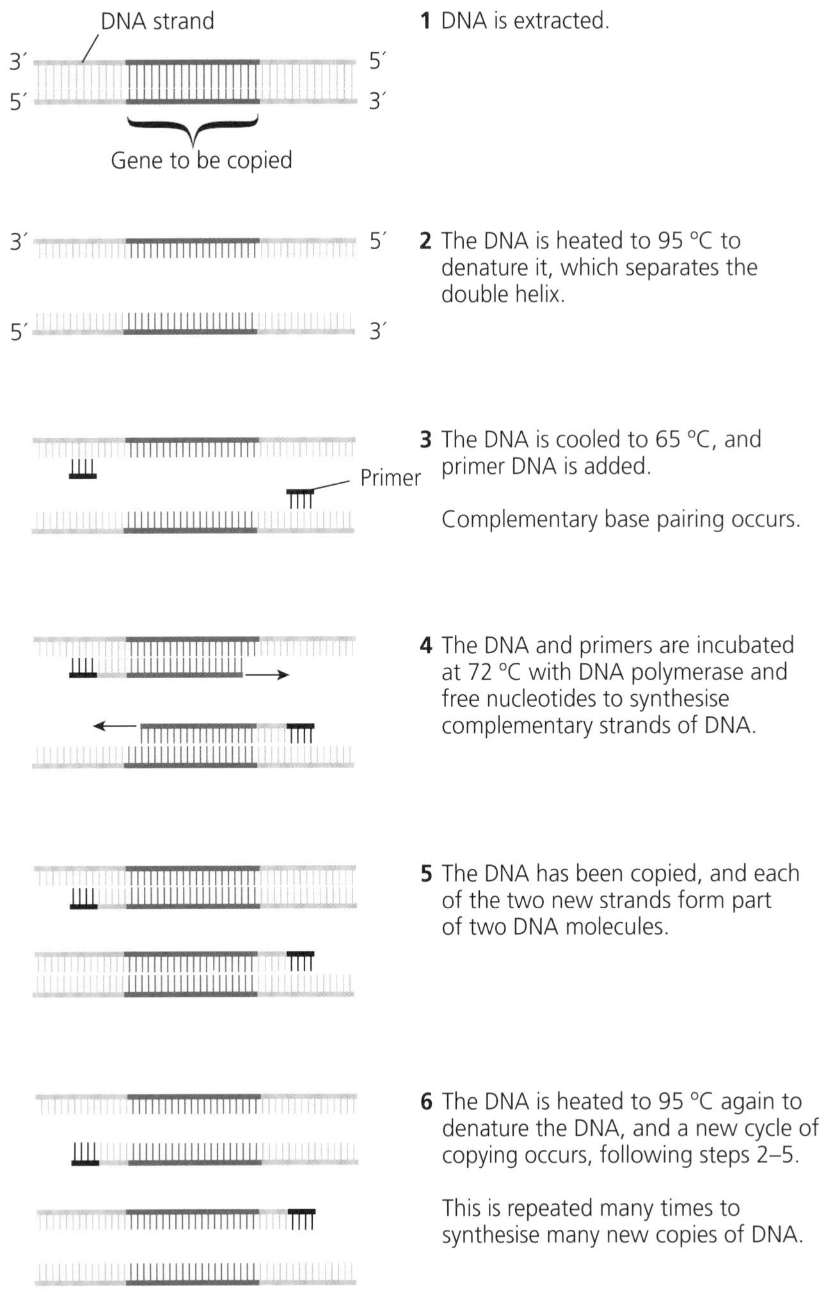

DNA strand

3′

5′

Gene to be copied

1 DNA is extracted.

3′ 5′

5′ 3′

2 The DNA is heated to 95 °C to denature it, which separates the double helix.

Primer

3 The DNA is cooled to 65 °C, and primer DNA is added.

Complementary base pairing occurs.

4 The DNA and primers are incubated at 72 °C with DNA polymerase and free nucleotides to synthesise complementary strands of DNA.

5 The DNA has been copied, and each of the two new strands form part of two DNA molecules.

6 The DNA is heated to 95 °C again to denature the DNA, and a new cycle of copying occurs, following steps 2–5.

This is repeated many times to synthesise many new copies of DNA.

Figure 19.1 The polymerase chain reaction

Before electrophoresis of DNA is carried out, the sample of DNA is exposed to a set of **restriction enzymes** (**restriction endonucleases**). These enzymes cut DNA molecules where particular base sequences are present. For example, a restriction enzyme called BamH1 cuts where the base sequence -GGATCC- is present on one strand of the DNA. Other restriction enzymes target different base sequences. This cuts the DNA into fragments of different lengths.

To carry out gel electrophoresis, a small, shallow tank is partly filled with a layer of agarose gel. A potential difference is applied across the gel, so that a direct current flows through it.

A mix of the DNA fragments to be separated is placed on the gel. DNA fragments carry a small negative charge, so they slowly move towards the positive terminal. The larger they are, the more slowly they move. After some time, the current is switched off and the DNA fragments stop moving through the gel.

The DNA fragments must be made visible in some way, so that their final positions can be determined. This can be done by adding fluorescent markers to the fragments. Alternatively, single strands of DNA made using radioactive isotopes, and with base sequences thought to be similar to those in the DNA fragments, can be added to the gel. These are called **probes**. They will pair up with fragments that have complementary base sequences, so their positions are now emitting radiation. This can be detected by its effects on a photographic plate (Figure 19.2).

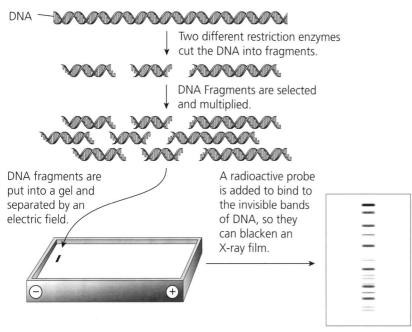

Figure 19.2 Gel electrophoresis

Applications of gel electrophoresis

Distinguishing between polypeptides or proteins

Polypeptides are made up of long chains of amino acids. These can differ in the charge they carry, because different R groups (p. 20) have different charges. For example, we have seen that sickle cell haemoglobin has an amino acid with an uncharged R group in the β polypeptide, whereas normal haemoglobin has an amino acid with a charged R group. This difference in charge means that the two different β globins will move at different speeds on the electrophoresis gel.

Distinguishing between different alleles of a gene

DNA strands containing different alleles of a gene may end up at different places on the gel after electrophoresis. For example, one allele may have more bases than another, and so be more massive and move more slowly.

Genetic fingerprinting

Some regions of human DNA are very variable, containing different numbers of repeated DNA sequences. These are known as variable number tandem repeats, or **VNTR**s. Each person's set of VNTR sequences is unique. Only identical twins share identical VNTR sequences.

Restriction enzymes are used to cut a DNA sample near VNTR regions. The chopped pieces of DNA are then separated using gel electrophoresis. Long VNTR sequences do not travel as far on the gel as short ones. The pattern of stripes produced is therefore determined by the particular combination of VNTRs that a person has.

Genetic fingerprinting can be used to determine whether:

- a sample of semen, blood or other tissue found at a crime scene could have come from the victim or a suspect
- a particular person could be the child, mother or father of another (Figure 19.3).

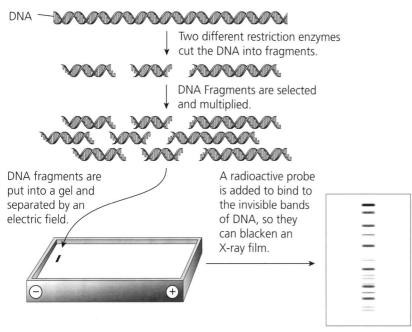

The dark areas on these autoradiographs of DNA represent particular DNA sequences.

In the first profile, you can see that all the bands on the child's results match up with either the mother's or potential father's, so this man could be the child's father.

However, on the second one, the child has a band that is not present in either the mother's or father's results. Some other person must therefore be the child's father.

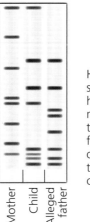

Figure 19.3 Electrophoresis in paternity testing

Using bacteria to make human insulin

In order to understand how genetic technology can produce recombinant organisms, we will look at the production of recombinant DNA containing the human insulin gene and its insertion into bacteria, which then express the gene and make insulin.

Insulin is a small protein. It is a hormone secreted by β cells in the islets of Langerhans in the pancreas in response to raised blood glucose concentration (p. 120).

In type I diabetes, no or insufficient insulin is secreted, and the person has to inject insulin. This used to be obtained from animals such as pigs. Today, almost all insulin used in this way is obtained from genetically modified bacteria.

Identifying the insulin gene
The amino acid sequence of insulin was already known. From this, the probable base sequence of the gene that codes for it, and of the mRNA transcribed from the gene, could be worked out (Figure 19.4).

Making the human insulin gene
Messenger RNA was extracted from β cells. These cells express the gene for insulin, so much of this mRNA had been transcribed from this gene. The appropriate mRNA was then incubated with the enzyme **reverse transcriptase**, which built single-stranded cDNA molecules against it. These were then converted to double-stranded DNA — the insulin gene (Figure 19.5).

Some extra single-stranded DNA was then added to each end of the DNA molecules. These are called **sticky ends**. Because they are single-stranded, they are able to form hydrogen bonds with other single-stranded DNA, enabling DNA molecules to join up with one another. This is important in a later stage of the process.

Cloning the DNA
Multiple copies of the DNA were then made (Figure 19.6) using **PCR** (p. 169).

Protein	Val	Asn	Gln	His	Leu	Cys

DNA
```
G T C A A T C A G C A C C T T T G T
C A G T T A G T C G T G G A A A C A
```

RNA C A G U U A G U C G U G G A A A C A

Figure 19.4

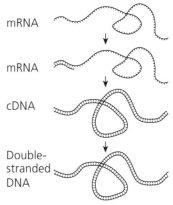

mRNA

mRNA

cDNA

Double-stranded DNA

Figure 19.5

Figure 19.6

Inserting the DNA into a plasmid vector

A **plasmid** is a small, circular DNA molecule found in many bacteria. A plasmid was cut open using a **restriction endonuclease**. The restriction endonucleases made a stepped cut across the DNA molecule, leaving single-stranded regions (Figure 19.7).

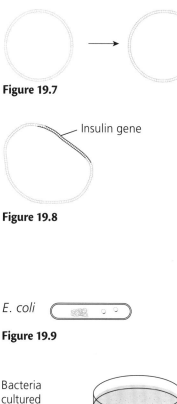

Figure 19.7

The cut plasmids and the insulin gene (the cloned DNA) were then mixed together, along with the enzyme **DNA ligase**. Complementary base pairing took place between the sticky ends added to the insulin genes and the sticky ends of the cut plasmids. DNA ligase then joined up the sugar–phosphate backbones of the DNA strands. This resulted in closed plasmids containing the insulin gene (Figure 19.8).

Insulin gene

Figure 19.8

Genes conferring resistance to an antibiotic were also introduced into the plasmids, next to the insulin gene.

Not all of the plasmids took up the gene. Some just closed back up again without it.

Inserting the plasmid into a bacterium

The plasmids were mixed with a culture of the bacterium *Escherichia coli*. About 1% of them took up the plasmids containing the insulin gene (Figure 19.9).

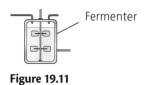

E. coli

Figure 19.9

Identifying the genetically modified bacteria

Antibiotics were then added to the culture of *E. coli* bacteria. The only ones that survived were the ones that had successfully taken up the plasmids containing the antibiotic resistance gene. Most of these plasmids would also have contained the insulin gene. Most of the surviving *E. coli* bacteria were therefore ones that now contained the human insulin gene (Figure 19.10).

Bacteria cultured on agar + antibiotic

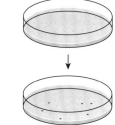

Surviving bacteria

Figure 19.10

Cloning the bacteria and harvesting the insulin

The bacteria were then grown in fermenters, where they were provided with nutrients and oxygen to allow them to reproduce to form large populations (Figure 19.11). Reproduction is asexual, so all the bacteria were genetically identical (clones).

These GM (genetically modified) bacteria are now cultured on a large scale. The bacteria synthesise and secrete insulin, which is harvested from the fermenters and purified before sale.

Fermenter

Figure 19.11

Vectors
Revised

In genetic technology, a **vector** is a structure that transfers the DNA into the organism that is to be modified. Vectors can be:

- plasmids (see above), which are suitable for use as vectors because:
  - they are small, so can easily be taken up by bacteria
  - their small size makes it easy to manipulate them without having large quantities of 'extra' DNA that are not required
  - they are made of a closed, circular molecule of DNA
- viruses, which are suitable for use as vectors because viruses naturally enter cells
- liposomes, which are small balls of lipid containing protein; being lipid, they are able to move easily through cell surface membranes and enter cells

Revision activity

- Summarise the uses of these enzymes in gene technology: DNA ligase, restriction endonuclease, DNA polymerase, reverse transcriptase.

Promoters

Revised

In bacteria, each gene is associated with a region of DNA called a **promoter** (p. 153). The enzyme **RNA polymerase** must bind to the promoter before it can begin transcribing the DNA to produce mRNA.

It was therefore important to ensure that there was a promoter associated with the human insulin gene when it was inserted into *E. coli*.

Markers

Revised

The antibiotic resistance genes added to the plasmids along with the human insulin gene act as **markers**. They make it possible to identify the bacteria that have taken up the gene.

There is concern that using antibiotic resistance genes as markers could increase the likelihood of the development of populations of harmful bacteria that are resistant to antibiotics. Today, the most common markers used are genes that code for the production of **fluorescent green protein**. The gene for this protein can be inserted along with the desired gene. Cells that fluoresce green are therefore likely to have taken up the desired gene.

Microarrays

Revised

A microarray is a tool used to identify particular DNA sequences in a sample of DNA. The microarray is a small piece of glass (often 2 cm^2) to which DNA probes (see p. 171) have been attached in a regular pattern. As many as 10 000 different probes can be attached to one microarray.

The DNA to be tested is cut into fragments using restriction endonucleases, and is then exposed to the microarray. Fragments whose base sequences are complementary to one of the probes will attach to it by complementary base pairing. The microarray can then be 'read' by a computer, which will identify the probes to which DNA fragments have attached, and therefore the base sequences present in the DNA sample.

This technique can be used to compare the **genome** (all the DNA present) in two different species, or in two individuals of the same species.

The technique can also be used to identify which genes are being expressed in a cell. mRNA is extracted from the cell, and then used to make cDNA (see p. 172). The cDNA is then analysed using a microarray. Any DNA that attaches to the probes must represent a gene that was being expressed and producing mRNA in the cell.

Genetic technology applied to medicine

Bioinformatics

Revised

Bioinformatics is the use of large databases and computer software to store and analyse information about living organisms. For example, there are now databases listing the base sequences of all the DNA in several individual people, as well as databases for the genomes of other species such as the nematode worm *Caenorhabditis* and the malarial parasite *Plasmodium*. There are also databases for the proteomes of different species. (The proteome is all the different proteins in an organism.)

The information stored in these databases is available to scientists all over the world. There are many potential uses for it. For example:

- People have different combinations of alleles of genes in their cells, and this can affect their susceptibility to diseases and the way in which drugs work in their bodies. It should one day be possibility to determine which is the best drug to give to a person with a particular DNA profile.
- Knowing the detailed DNA sequences of different species or varieties of *Plasmodium* can help scientists to develop drugs that will act against them.
- Matches can be made between DNA sequences and protein sequences, which can help to unravel how genes affect metabolism and therefore health.
- Models can be made of enzymes using amino acid sequences derived from the proteome, and drugs can then be designed that could inhibit these enzymes, using computer modelling.

Advantages of producing human proteins by gene technology   Revised

Examples of human proteins produced by GM organisms include:

- insulin for the treatment of diabetes, produced by *E. coli* (see pp. 172–173)
- factor VIII for the treatment of haemophilia, produced by tissue cultures of GM hamster cells
- adenosine deaminase for the treatment of severe combined immunodeficiency (SCID), produced by GM bacteria

There are numerous advantages in using proteins produced in this way. For example:

- The insulin produced by the genetically engineered *E. coli* is identical to human insulin, because it is made by following the genetic code on the human insulin gene. Insulin obtained from the pancreas of an animal is slightly different, and therefore may have different effects when used to treat diabetes in humans.
- Large quantities of insulin can be made continuously using *E. coli*, and this can be done under controlled conditions. Only small quantities of insulin can be obtained from the pancreas of an animal, and it is not easy to purify the insulin to produce a standard product that is safe for medicinal use.
- Many religions and cultures, and also many individuals, are against the idea of harvesting insulin from a dead animal for use in humans.

Genetic screening   Revised

Finding out the genes that a person has is called **genetic screening**. This can be used

- to identify people who are carriers, i.e. who have a copy of a harmful recessive allele, such as the cystic fibrosis allele, the sickle cell anaemia allele or the haemophilia allele; a couple with one of these genetic conditions in the family could therefore find out if they are both heterozygous and therefore might have a child with the condition
- in preimplantation genetic diagnosis, to check the genes of an embryo produced in vitro (i.e. by fertilisation outside the body) before it is placed in the mother's uterus; this can ensure that only embryos that do not have the genes for a genetic disease are implanted

- for prenatal testing, i.e. checking the genes of an embryo or fetus in the uterus; this could enable the mother to decide to have her pregnancy terminated if the baby would have a genetic disease
- to identify people who will develop a genetic condition later in life; for example, Huntington's disease is caused by a dominant allele, but does not manifest itself until middle age; a person with this disease in the family could check if they have the gene before they decide to have children themselves;
- to identify people with alleles that put them at risk of developing other diseases; for example, a woman who has relatives with breast cancer could find out if she has the *BRCA1* or *BRCA2* alleles, which are known to be associated with an increased risk of breast cancer, and could decide to have a mastectomy to prevent developing the illness

A **genetic counsellor** helps people to interpret the results of the screening and therefore to make decisions. This can be very difficult, and involves moral and ethical, as well as scientific, issues. For example, if a pregnant woman finds that her child will have cystic fibrosis, are these sufficient grounds to have her pregnancy terminated, or does the child have a right to life?

Treating genetic conditions using gene therapy   Revised

Gene therapy is the treatment of a genetic disease by changing the genes in a person's cells. It is only suitable for:

- diseases caused by a single gene
- diseases caused by a recessive allele of a gene
- serious diseases for which treatment is limited and no other cure is possible

Although attempts have been made to treat several different diseases using gene therapy, there are still many problems to be solved before treatments become widely available and successful.

The vectors used for gene therapy can be liposomes or viruses. In a few cases, trials have been carried out using 'naked' DNA.

Gene therapy for cystic fibrosis

Cystic fibrosis (CF) is a genetic condition resulting from a mutation in a gene that codes for a transporter protein called **CFTR**. This protein lies in the cell surface membrane of cells in many parts of the body, including the lungs, pancreas and reproductive organs. The CFTR protein actively transports chloride ions out of cells.

There are several different mutations that result in changes in the CFTR protein. Some of them involve a change in just one base in the gene coding for the CFTR protein. The commonest one, however, involves the loss of three bases from the gene, meaning that one amino acid is missed out when the CFTR protein is being made. In all cases, the protein that is made does not work properly.

Normally, chloride ions are transported out of the cells through the CFTR protein. Water follows by osmosis. When the CFTR protein is not working, this does not happen. There is therefore less water on the outer surface of the cells than there should be. The mucus that is produced in these areas therefore does not mix with water in the usual way. The mucus is thick and sticky. As a result:

- the abnormally thick mucus collects in the lungs, interfering with gas exchange and increasing the chance of bacterial infections
- the pancreatic duct may also become blocked with sticky mucus, interfering with digestion in the small intestine
- reproductive passages, such as the vas deferens, may become blocked, making a person sterile

Attempts have been made to treat cystic fibrosis by introducing the normal *CFTR* allele into a person's cells. Two methods have been trialled:

- inserting the normal gene into a harmless virus and then allowing the virus to infect cells in the person's respiratory passages — the virus enters the cells and introduces the gene to them
- inserting the gene into little balls of lipid and protein, called liposomes, and spraying these as an aerosol into a person's respiratory passages

The virus and the liposomes are said to be **vectors** — they transfer the gene into the person's cells.

In each case there was some success, in that some of the cells lining the respiratory passages did take up the gene. Because the normal gene is dominant, there only needs to be one copy in a cell for it to produce normal mucus. There is no need to remove the faulty allele first, because it is recessive.

However, there were problems with the trials of gene therapy for cystic fibrosis:

- Only a few cells took up the normal gene, so only these cells produced normal mucus.
- It was only possible for cells in the respiratory passages to take up the normal gene, not cells in the pancreas or reproductive organs.
- Cells in the surfaces of the respiratory passages do not live for very long, so treatment would need to be repeated every few weeks.
- When using the virus as a vector, some people developed serious lung infections.

Gene therapy for other conditions

SCID

SCID is caused by a lack of the enzyme adenosine deaminase. This results from mutations in the gene for this enzyme, which result in it not being expressed. Trials have been carried out using viruses as vectors to transfer the normal allele into white blood cells taken from a person with SCID. These trials have been partly successful, with many patients now having a functional immune system. However, there is no control over where the allele is inserted into the patient's DNA, and in some case this has resulted in genes controlling cell division becoming altered and causing cancer (leukaemia). Different methods of transferring the gene are now being trialled.

Choroideremia

This is a genetic condition in which cells in the retina gradually die, and is due to a faulty allele of the *CHM* gene on the X chromosome. Several patients have now had their eyesight partially restored after viruses carrying normal alleles of the *CHM* gene were injected beneath their retinas in one eye.

Social and ethical considerations Revised

The ability to detect genetic conditions, and perhaps to cure some of them, raises ethical questions that society needs to debate. You might like to think about some of the questions raised by the examples given below, and consider the answers to those questions. There are no 'right' or 'wrong' answers. Try to think about two different viewpoints for each one.

- A person undergoes genetic screening and finds that they have a genetic condition that cannot be treated. How does this affect their life?
- A couple undergo genetic screening and find that they both have a recessive allele for an unpleasant but not life-threatening genetic condition. What do they do?

- A mother finds that her embryo will have cystic fibrosis. Does she have her pregnancy terminated?
- A person with a genetic condition that would have prevented them living to reproductive age has gene therapy, and then has children who have inherited the faulty allele. What effects might this have on the population?
- A couple very much want a girl baby rather than a boy. They decide to have IVF and have the sperm screened so that they can choose a sperm carrying an X chromosome to fertilise the egg. Should this be allowed?

Now test yourself

2 Explain why gene therapy is most likely to succeed in treating conditions caused by a recessive allele of a single gene.

Answers on p.206

Tested

Genetically modified organisms in agriculture

We have seen that crops can be improved by selective breeding. However, this takes a long time, and can only involve alleles that are already present in the genome of that crop species. Genetic engineering allows the targeted addition of alleles for a desired characteristic into a species where that allele is not naturally present. Although it is a highly technical and expensive process, it can achieve results more quickly than conventional selective breeding.

Examples of GM crops and livestock

Revised

Golden Rice

This is rice that has had genes encoding vitamin A added to it. In countries where rice forms a major part of the diet, children can suffer from vitamin A deficiency, causing blindness and severe deficiencies of the immune system, which can result in death. Eating this GM rice provides more β carotene, which is used to make vitamin A, and so can help to avoid this condition.

The company and researchers who developed the rice have donated it free for use in developing countries. The GM rice has been interbred with locally developed varieties of rice, so that it will grow well in different countries.

However, as yet no government has allowed it to be grown by farmers. Some people argue that this is not the best way to solve problems associated with poverty, and that it would be better to improve people's lives so that they are able to choose a wider variety of foods to eat. They are also concerned that the GM rice might somehow be harmful to health, although there is no scientific foundation for such concerns.

Insect-resistant maize, cotton and tobacco

Yields of maize can be greatly reduced by an insect larva called the corn borer. Cotton yields are reduced by the cotton bollworm, which is also an insect larva. Tobacco plant yields are affected by the tobacco budworm.

Pesticides sprayed onto the crops can kill these insects. However, the pesticides can also harm other, beneficial insects. The insects also evolve resistance to the pesticides.

Genes that code for the production of a protein derived from the bacterium *Bacillus thuringiensis*, called Bt toxin, have been inserted into maize, cotton and tobacco plants. The plants therefore produce the protein, which is converted into the toxin once inside the gut of insects that have eaten the leaves. This means that the toxin kills insects that feed on the plants, but not other insects.

Benefits include the following:

- There is less loss of the crop to insect pests, so greater yields are obtained.

- Less or no insecticide needs to be sprayed on the crop, reducing harm to non-target insect species.
- Only insects that eat the plant are harmed, not other insects, which are affected when insecticides are sprayed on crops.
- It is less likely that insect pests will evolve resistance to the Bt toxin than to pesticides. However, there are signs that resistance can develop in some pest species. This can be counteracted by using slightly different forms of the Bt toxin, or a combination of two different Bt toxins, in GM crops.

Possible detrimental effects include the following:

- The seed of Bt corn and cotton costs more for farmers to buy than non-GM seed, making it difficult for farmers in developing countries to afford it, and therefore making it difficult for them to compete with farmers in richer countries.
- It is possible that some non-pest insects might be harmed by the Bt toxin. However, overall it is found that there are more non-pest insects present in fields where Bt corn or Bt cotton are grown than in fields where non-Bt crops are grown and sprayed with pesticides.
- Genes from the Bt crops might be transferred (e.g. in pollen) to species of wild plants growing nearby. However, no evidence has been seen of this happening to a significant degree.

Herbicide-resistant oilseed rape

Oilseed rape, *Brassica napus*, is grown for the oil that can be extracted from its seeds. This is used in food production, and also to produce biofuels and lubricants. GM varieties of oilseed rape contain genes that make the plants resistant to the herbicide glyphosate. This means that farmers can spray the field with herbicide, which will kill weeds but not the crop. Growing GM oilseed rape therefore increases yields and reduces costs. A possible detrimental effect is that the herbicide resistance genes could be transferred to species of wild plants growing nearby, but so far little evidence of this has been detected.

GM salmon

Salmon are farmed for food in many parts of the world. A company in the USA has developed a GM variety of salmon that contains a growth hormone gene from another species of salmon, and a promoter from a different species of fish. This makes the GM salmon grow much faster and larger than normal. This could allow more food to be produced more quickly. However, there are concerns that, if these fish escape from the cages where they are kept, they could reproduce in the wild and adversely affect populations of other marine organisms. To avoid these problems, it is planned to farm only sterile female fish. Moreover, the GM salmon are all triploid, making it difficult for them to produce viable gametes. It is also recommended that the fish are only grown in land-based water tanks, so that they are much less likely to escape into the wild.

A level experimental skills and investigations

Skills and mark allocations

More than one quarter of the marks in your A level examinations are for practical skills. These are assessed in a written paper — *not* by a practical examination. However, you will need to do plenty of practical work in a laboratory throughout your A level biology course in order to develop the practical skills that are assessed.

There is a total of 30 marks available on this paper. Although the questions are different on each practical paper, the number of marks assigned to each skill is always the same. This is shown in the table below.

Skill	Total marks	Breakdown of marks	
Planning	15 marks	Defining the problem	5 marks
		Methods	10 marks
Analysis, conclusions and evaluation	15 marks	Dealing with data	8 marks
		Evaluation	4 marks
		Conclusion	3 marks

The syllabus explains each of these skills in detail, and it is important that you read the appropriate pages in the syllabus so that you know what each skill is, and what you will be tested on.

The next few pages explain what you can do to make sure you get as many marks as possible for each of these skills. They have been arranged by the kind of task you will be asked to do.

Some of the skills are the same as for AS. However, most of them are now a little more demanding. For example, in dealing with data, you are not only expected to be able to draw results charts and graphs, but also to use statistics to determine what your results show.

How to make the most of your practical skills

Planning
Revised

At least one of the questions on the practical paper will require you to plan an investigation.

The question will describe a particular situation to you, and ask you to design an experimental investigation to test a particular hypothesis or prediction. The planning question will usually involve investigating the effect of one factor (the independent variable) on another (the dependent variable).

Sometimes, the question will tell you the apparatus you should use. In other cases, you will need to decide on the apparatus. If you have done all of the practicals listed in the syllabus during your course, then you will have had direct experience of a wide range of apparatus and techniques, which will make this much easier for you.

Stating a hypothesis

You may be asked to state a hypothesis. For example, you might be asked to plan an experiment to find out how light intensity affects the rate of photosynthesis. Your hypothesis could be:

Increasing light intensity will increase the rate of photosynthesis.

Notice that your hypothesis should include:
- reference to the independent variable (light intensity)
- reference to the dependent variable (rate of photosynthesis)
- a clear statement of how a particular change in the independent variable (an increase) will affect the dependent variable (an increase)

You could be asked to express your hypothesis in the form of a sketch graph. Here, you could draw a graph like this:

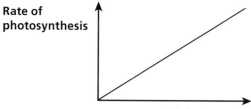

Your hypothesis should be:
- **quantifiable** — this means that you should be able to *measure* changes in the independent and dependent variables and obtain numerical results
- **testable** — this means that you should be able to do an experiment that will test whether or not your hypothesis could be correct
- **falsifiable** — this means that you should be able to do an experiment whose results could show that your hypothesis is not correct

Notice that it is not possible to actually *prove* that a hypothesis is correct. For example, if your experiment involved putting a piece of pondweed in a tube and measuring the rate at which oxygen is given off with a lamp at different distances from the tube, you might find that the rate of photosynthesis does indeed increase with light intensity. You could say that your results 'support' your hypothesis. However, you cannot say that they 'prove' it, because you would need to do more experiments and obtain a lot more data before you could be sure this relationship is *always* true.

However, you can *disprove* a hypothesis. If you kept on increasing the light intensity in this situation, you would probably find that the rate of photosynthesis eventually levels off, as some other factor begins to limit the rate. This would disprove your hypothesis.

Variables

You will usually need to decide what are the *independent* and *dependent variables*. You will already have thought about this for AS, and it is described on p. 77.

You will also need to identify important *variables that should be controlled* or standardised. This is described on p. 79.

Range and intervals

You will need to specify a suitable range and suitable intervals for the independent variable that you would use. This is described on p. 77.

> **Expert tip**
>
> Check exactly what the question is asking you do. You may be expected to state your hypothesis both in words and in the form of a predicted graph.

> **Expert tip**
>
> Even if you are not asked to write them down, make sure that you have identified the independent and dependent variables. State clearly which variables you would control, and be prepared to state exactly how you would do this.

Apparatus

Sometimes you will be told exactly what apparatus you should use. If so, then it is important that your experiment does use this apparatus. If you describe a different kind of apparatus, you will not get many marks.

For example, the question may include a diagram of a plant shoot in a test tube containing dye solution, with a layer of oil on the liquid surface. The question may ask how you could use this apparatus to find the rate of movement of water up the stem. You will probably never have used this apparatus for this purpose, but you should be able to work out a way of using it for this investigation. If you describe the use of a different piece of apparatus — for example, a potometer — then you will not get many marks, because you have not answered the question.

If you do plenty of practical work during your biology course, then you will become familiar with a wide range of apparatus. There is a list of basic apparatus that you should be able to use in the syllabus, under the heading 'Laboratory equipment'.

> **Expert tip**
>
> Take care that your plan uses the apparatus that the question asks you to use, even if you think this is not the best apparatus, or if it is not apparatus that you have used before.

Method

You will need to describe exactly what you will do in your experiment.

Changing and measuring the independent variable

You should be able to describe *how* you will change the independent variable and *how* you will measure it. If the independent variable is the concentration of a solution, then you should be able to describe how to make up a solution of a particular concentration.

To make up a solution of $1 \, \text{mol dm}^{-3}$, you would place 1 mole of the solute in a $1 \, \text{dm}^3$ flask, and then add a small volume of solvent (usually water). Mix until fully dissolved, then add solvent to make up to exactly $1 \, \text{dm}^3$. (1 mole is the relative molecular mass of a substance in grams.)

To make up a 1% solution, you would place 1 g of the solute in a container with a small volume of solvent. Mix until fully dissolved, then make up to $100 \, \text{cm}^3$ with solvent.

You should be able to use a stock solution to make serial dilutions. This is described on p. 78. It is worth remembering that a small test-tube holds about $15 \, \text{cm}^3$, and a large one (sometimes called a boiling tube) up to $30 \, \text{cm}^3$.

Keeping key variables constant

You will be expected to describe *how* you will control a particular variable. For example, if you need to control temperature, then you should explain that you would use a water bath (either thermostatically controlled, or a beaker of water over a Bunsen burner) in which you take the temperature using a thermometer.

Measuring the dependent variable

You should describe *how* you will measure the dependent variable, and *when* you will measure it.

Sequence of steps

Your method should fully describe the sequence of the steps you will carry out. This will include setting up the apparatus and all the other points described above relating to changing, measuring and controlling variables. You may like to set this out as a series of numbered points.

Risk assessment

You should be able to identify any ways in which your experiment might cause a risk of injury or accident, and to suggest safety precautions related to these risks. Sometimes these are obvious — for example, if you are using concentrated acid, then there is a risk that this might come into contact with clothing, skin or eyes, and therefore protective clothing and eye protection should be worn. In the experiment investigating the effect of light intensity on the rate of photosynthesis, care should be taken not to get water on the electrical light source, as this could cause a current surge or give a person an electric shock.

Sometimes, an experiment might be so simple and safe that there are no genuine risks. If so, then you should say so. Do not invent risks if there are none!

Expert tip

Always think about risk. Identify any genuine risks and explain how you would minimise them. Do not invent risks if there are none.

Reliability
As described on p. 82, it is a good idea to do at least *three repeats* for each value of the independent variable. A mean can then be calculated, which is more likely to give a true value than any one of the individual values.

Expert tip

It is almost always a good idea to suggest doing at least three repeats.

Recording and displaying data, and drawing conclusions
This is the same as for AS. It is described on pp. 81–87.

Dealing with data
Revised

Tables and graphs
You may be asked to draw tables or graphs to record and display data. These are described on pp. 81–86.

Calculations to summarise or describe data
You will be asked to carry out some type of calculation, using data that you have been given.

Mean, median and mode
These are best explained using an example. Let's say that you had measured the lengths of 20 leaves. Your results, recorded in mm, were:

33.5, 56.5, 62.0, 75.0, 36.0, 54.5, 43.5, 41.5, 54.0, 53.0

53.5, 39.0, 72.5, 66.5, 58.5, 42.5, 41.5, 49.0, 69.0, 38.5

(Notice that all the measurements are to three significant figures, or one decimal place, and that all values have been measured to the nearest 0.5 mm.)

To calculate the **mean** value, add up all the values and divide by the total number of measurements.

$$\text{mean value} = \frac{\text{sum of individual values}}{\text{number of values}} = \frac{1040}{20} = 52.0$$

Now let's say you measured the lengths of 80 more leaves, and obtained the results displayed in this histogram.

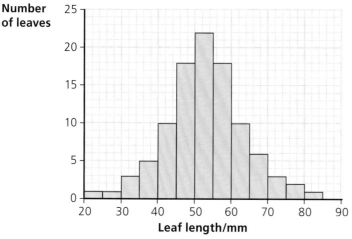

A frequency histogram

The **mode**, or **modal class**, is the most common value in the set of observations. In this instance, this is 50–54 mm. The **median** is the middle value of all the values. In this case, the median value is 52.5 mm.

If the frequency histogram of your data is a perfect bell-curve, we say that the data show a **normal distribution**. For a perfect normal distribution, the mean, mode and median are all the same.

Percentage increase or decrease

You can remind yourself how to calculate these on p. 82.

Range and inter-quartile range

You have already met the idea of 'range', when thinking about the range of values you would use for an independent variable in an investigation. The **range** is the spread between the highest and lowest values in your data. For the 100 leaves in the histogram above, the range is 65 mm.

The **inter-quartile range** describes the range between the values that are one quarter and three quarters along the complete range of your data. For the leaf data, the inter-quartile range is from 36.25 to 68.75. These values tell you how much the data are spread. If they are all very close to each other, then the range and the inter-quartile range will be small.

This can be useful if you are looking at a set of repeat results in an experiment. Ideally, you would hope that all the results are very close to one another. If they are not, then this indicates a lack of reliability — there is a lot of variation in your results for a particular value of your independent variable and therefore you cannot have a great deal of faith that your mean is the 'true' value.

Statistics

Statistical methods are designed to help us to decide the level of confidence we can have in what our results seem to be telling us. This is often a problem in biology, where we are generally dealing with variable organisms and may not be able to fully control all the control variables we would like. This means that we do not get a set of results that precisely matches what we have predicted from our hypothesis, or we are not really sure whether the set of results we have obtained for one experiment is genuinely different from the set of results we have obtained for another experiment.

In most statistical tests, you will need to calculate some basic values along the way. These include **standard deviation** and **standard error**.

Calculating standard deviation

This is used with data that show an approximately normal distribution. It is a measure of how much the data vary around the mean.

The larger the standard deviation, the wider the variation.

This is how you would calculate the standard variation for the leaf lengths shown on p. 183. These were:

> 33.5, 56.5, 62.0, 75.0, 36.0, 54.5, 43.5, 41.5, 54.0, 53.0
>
> 53.5, 39.0, 72.5, 66.5, 58.5, 42.5, 41.5, 49.0, 69.0, 38.5

1 Calculate the mean. This is 52.0.
2 For each measurement, calculate how much it differs from the mean, and then square each of these values. It is easiest to do this if you set the results out in a table. In the table:
   - x represents an individual value
   - $\overline{X}$ represents the mean
   - Σ represents 'sum of'

x	$(x - \bar{X})$	$(x - \bar{X})^2$
33.5	−18.5	342.25
56.5	4.5	20.25
62.0	10.0	100.00
75.0	23.0	529.00
36.0	−16.0	256.00
54.5	2.5	6.25
43.5	−8.5	72.25
41.5	−10.5	110.25
54.0	2.0	4.00
53.0	1.0	1.00
53.5	1.5	2.25
39.0	−13.0	169.00
72.5	20.5	420.25
66.5	14.5	210.25
58.5	6.5	42.25
42.5	−9.5	90.25
41.5	−10.5	110.25
49.0	−3.0	9.00
69.0	17.0	289.00
38.5	−13.5	182.25
		$\Sigma(x - \bar{X})^2 = 2966$

3 Now you need to use the following formula. You do *not* need to memorise this formula, because if you are asked to use it in the exam, you will always be given it.

$$s = \sqrt{\frac{\Sigma(x - \bar{X})^2}{n - 1}}$$

where:

s is the standard deviation

n is the number of readings (which in this case was 20)

So:

$$s = \sqrt{\frac{2966}{19}} = 12.49$$

This should be rounded up to 12.5.

All the values used in the calculations are in mm, so the standard deviation is also in mm. You can say that the mean length of the leaves in this sample is 52.0 mm, with a standard deviation of 12.5. This is often written as: 52.0 mm ± 12.5 mm.

Calculating standard error

The set of data about leaf lengths was taken from a *sample* of leaves. How likely is it that the mean of these data is actually the true mean for the entire population? We can get some idea of this by calculating the standard error for these data. The standard error tells us how far away the actual mean for the entire population might be from the value we have calculated for our sample.

The formula for standard error, S_M, is:

$$S_M = \frac{s}{\sqrt{n}}$$

For the leaf data, this works out as:

$$\frac{12.5}{\sqrt{20}} = \frac{12.5}{4.5} = 2.8 \, \text{mm}$$

This tells us that, if we took a fresh sample of leaves from the same population, we could be 95% confident that the mean length would be within 2 × 2.8 mm of our original mean, which was 52.0 mm.

We could use this to show 95% confidence limits on a graph. Let's say you were trying to find out if the lengths of leaves from a species of tree growing at the edge of a wood were different from the lengths of leaves from the same species growing in the middle of a wood. The first set of data (the ones you have been looking at already) came from the edge of the wood. Here are the results you got when you measured 20 leaves from the middle of the wood.

Now test yourself

1 Show that the standard deviation for the leaves from the middle of the wood is 10.4 mm.

2 Show that the standard error for these data is 2.3 mm.

Answers on pp.206–207

Tested

23.0, 46.5, 52.0, 32.5, 43.0, 50.5, 26.5, 31.5, 54.0, 28.0

47.5, 39.0, 58.5, 47.5, 33.5, 42.5, 26.5, 51.0, 38.0, 38.5

The mean of these values is 40.5 mm.

We can then calculate the standard deviation, which is 10.4 mm. The standard error is 2.3 mm.

The values for standard error can be shown as error bars on a bar chart.

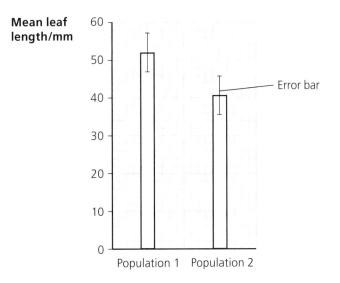

If the values for the two error bars overlap, this means there is not a significant difference between the mean lengths of the two populations of leaves. Here, there is no overlap, so we can say that it is possible that there is a significant difference between the lengths of the leaves from the middle and the edge of the wood.

The chi-squared test

This test is used to tell you whether any difference between your observed and expected values is due to chance, or whether it means that your null hypothesis cannot be correct. It can also be used to test for whether there is an association between two variables — for example, whether people who are left-handed are more likely to study physics A level than people who are right-handed. The chi-squared test is described on pp. 150–151.

The t-test

This test is used to tell you whether the means of two sets of values, each taken from a population of data that has a normal distribution, are significantly different from one another.

We can use the leaf length data to illustrate how this test is done, and what it tells us.

Here are the two sets of data again:

leaves from the edge of the wood

> 33.5, 56.5, 62.0, 75.0, 36.0, 54.5, 43.5, 41.5, 54.0, 53.0

> 53.5, 39.0, 72.5, 66.5, 58.5, 42.5, 41.5, 49.0, 69.0, 38.5

leaves from the middle of the wood

> 23.0, 46.5, 52.0, 32.5, 43.0, 50.5, 26.5, 31.5, 54.0, 28.0

> 47.5, 39.0, 58.5, 47.5, 33.5, 42.5, 26.5, 51.0, 38.0, 38.5

Just as for the chi-squared test, you begin by constructing a null hypothesis. This would be:

There is no significant difference between the means of the lengths of the leaves from the edge of the wood and from the middle of the wood.

The formula for the *t*-test will always be given to you, so you do not need to learn it. It is:

$$t = \frac{|\bar{x}_1 - \bar{x}_2|}{\sqrt{\frac{s_1^2}{n_1} + \frac{s_2^2}{n_2}}}$$

The symbols all mean the same as before. We have already calculated these values. They are as follows:

$\bar{x}_1$ is the mean for the first set of leaves, 52.0 mm.

$\bar{x}_2$ is the mean for the second set of leaves, 40.5 mm.

s_1 is the standard deviation for the first set of leaves, 12.5 mm.

s_2 is the standard deviation for the second set of leaves, 10.4 mm.

n_1 and n_2 are the numbers in each sample, which were both 20.

Substituting these values into the formula, we get:

$$t = \frac{|52 - 40.5|}{\sqrt{\frac{12.5^2}{20} + \frac{10.4^2}{20}}}$$

$$t = \frac{11.5}{\sqrt{\frac{156.25}{20} + \frac{108.16}{20}}}$$

$$t = 3.16$$

What does this mean? You now have to look up this value of *t* in a table of probabilities.

First, you need to decide how many degrees of freedom there are in your data. To do this, use the formula:

> degrees of freedom = $n_1 + n_2 - 2$
>
> $= 20 + 20 - 2$

So the number of degrees of freedom is 38.

Here is a small part of the table showing values of *t*.

Degrees of freedom	Probability that null hypothesis is correct			
	0.10	0.05	0.02	0.01
25	1.70	2.06	2.49	2.79
30	1.70	2.04	2.46	2.75
40	1.68	2.02	2.42	2.70

Your value of t is 3.16. The table does not have a row for 38 degrees of freedom, so we can look at the closest one, which is 40. There is no value in this row that is as high as your value for t. The values in the table are increasing from left to right, so we can say that our value of t would lie well to the right of anything in the table. This means that the probability of the null hypothesis being correct is much less than 0.01.

Just as we did for the chi-squared test (pp. 150–151), we take a probability of 0.05 as being the decider for the t-test. If the probability is lower than this, then we can say it is very unlikely that the null hypothesis is correct. In other words, there *is* a significant difference between the mean lengths of the two populations of leaves. The mean length of the leaves in the middle of the wood is significantly less than the mean length of the leaves at the edge of the wood.

Spearman's rank correlation

This test is used to test for a correlation between two sets of data that are not distributed normally. You can use this test, for example, to see if there is a correlation between the distribution and abundance of two species in a habitat. In order to use the test, you should have:

- two sets of data where the data points are independent of each other
- ordinal data — that is, data that consist of numerical scores that can be ranked
- drawn a scatter diagram of one set of data against the other, and it looks as though they might be correlated
- ideally, between 10 and 30 pairs of observations
- data that were collected randomly within the population

The formula for Spearman's rank correlation, r_s, is:

$$r_s = 1 - \frac{6 \times \Sigma D^2}{n^3 - n}$$

where n is the number of pairs of observations, and D is the difference between each pair of ranked measurements.

For example, imagine you have used quadrats to sample two species, X and Y, on a rocky shore. You have counted the number of each species in each quadrat. Here are your results:

Quadrat	Number of species X	Number of species Y
1	14	33
2	8	4
3	22	41
4	1	0
5	16	20
6	5	4
7	12	23
8	7	16
9	9	21
10	19	38

You think that your data suggest that there might be a correlation between the abundance of the two species — i.e. high numbers of species X appear to be associated with high numbers of species Y.

First, rank the data for each of the species. The largest number has rank of 1:

Quadrat	Number of species X	Rank for species X	Number of species Y	Rank for species Y
1	14	4	33	3
2	8	7	4	8=
3	22	1	41	1
4	1	10	0	10
5	16	3	20	6
6	5	9	4	8=
7	12	5	23	4
8	7	8	16	7
9	9	6	21	5
10	19	2	38	2

Now calculate the differences in rank between the two species, and square it:

Quadrat	Rank for species X	Rank for species Y	Difference in rank, D	D^2
1	4	3	1	1
2	7	8=	−1	1
3	1	1	0	0
4	10	10	0	0
5	3	6	3	9
6	9	8=	1	1
7	5	4	1	1
8	8	7	1	1
9	6	5	1	1
10	2	2	0	0
				$\Sigma D^2 = 15$

Substitute into the equation:

$$r_s = 1 - \frac{6 \times 15}{10^3 - 10}$$

$$= 1 - \frac{90}{990}$$

$$= 1 - 0.09$$

$$= 0.91$$

The closer this number is to 1, the greater the degree of correlation between the two sets of data. You may be given a table like this, against which you can compare your calculated value, to see if the correlation is significant. If your value for r_s is equal to or greater than the number in the table, then there is a significant correlation.

n	5	6	7	8	9	10	11	12	14	16
Critical value of r_s	1.00	0.89	0.79	0.76	0.68	0.65	0.60	0.54	0.51	0.51

For these two sets of data, n is 10 and r_s is 0.91, which is much greater than the critical value of 0.65. So we can say that that there is a correlation between the abundances of the two species.

Now test yourself

3 A student recorded the numbers of species P and species Q in 10 quadrats. The results are shown in the list. In each case, the first number is the number of species P, and the second number is species Q.

   6, 5; 12, 22; 1, 3; 0, 4; 16, 35; 8, 10; 11, 11; 4, 9; 5, 3; 12, 21

   Use Spearman's rank correlation to determine whether there is a correlation between the distribution of species P and species Q.

Answers on p.207

Tested

Pearson's linear correlation

This test is used when you think there may be a linear correlation between two sets of normally distributed data. You can use it when:

- you have two sets of continuous data — for example, reaction time and volume of coffee drunk; numbers of species A and numbers of species B; mass of seed and time it takes to fall to the ground
- you have drawn a graph with one set of data on the x-axis and the other set on the y-axis, and it looks as though you could draw a straight line through the points
- the data have come from a population that is normally distributed
- you have at least five sets of paired data

The formula for Pearson's linear correlation, r, is:

$$r = \frac{\Sigma xy - n\bar{x}\,\bar{y}}{n\,s_x s_y}$$

where:

x and y are the individual values of your results (observations)

$\bar{x}$ and $\bar{y}$ are the means of the two sets of observations

n is the number of observations

s_x and s_y are the standard deviations for x and y

If your calculation gives you $r = 1$ or -1, then the correlation is perfectly linear. This means that, if you plotted one set of data on the x-axis and the other set on the y-axis of a graph, every point would lie exactly on a straight line.

These are the steps you need to follow:

1 Draw a table and fill in all your results for x and y, with 'matching' pairs opposite one another, just like the first one for Spearman's rank correlation on p. 188.
2 Calculate the value of xy for each pair of data, by multiplying each pair. You can do this in an extra column on the right of your table.
3 Add up all your values of xy to find Σxy.
4 Calculate the means, $\bar{x}$ and $\bar{y}$.
5 Calculate $n\bar{x}\,\bar{y}$.
6 Calculate the standard deviation for each set of data (see p. 185). (You may have a calculator that can do this quickly, or you can find a website that does it for you — you just have to key in all of your data and press a button.)
7 Now substitute all of your numbers into the formula above.

> **Expert tip**
>
> Note that, even if Spearman's rank correlation or Pearson's linear correlation test shows that there is a significant correlation between your two sets of data, this does not mean that one is causing the other. Correlation does not necessarily imply a causal relationship.

Now test yourself

Tested

4 A student measured the length of the shell of 10 molluscs, and their body mass. These are her results:

Shell length/mm	Body mass/g
38	6.1
18	3.6
20	3.2
31	5.7
12	2.6
12	3.2
25	4.7
20	3.6
29	4.7
19	3.3

Use Pearson's linear correlation test to determine whether there is a linear correlation between shell length and body mass.

Answer on p.207

Evaluation

This skill builds on what you have already been doing at AS. You need to be able to:

- **spot anomalous data**. These are values that do not fit into a strong pattern that is shown by the rest of the data. This may indicate that you made a mistake when taking a measurement, or that some other variable had changed significantly for that one reading. The best thing to do with genuinely anomalous data is to take them out of your results. Do not include them in calculations of means, and do not take any notice of them when deciding where to draw a line on a graph.

- **consider the need for repeats or replicates**. If you have decided that a data point is anomalous, then it would be a good idea to take repeat readings for that value. Only by repeating those readings can you be sure that your odd result really is anomalous. In any case, it is often a good idea to include a set of repeats or replicates, as explained on pp. 82–83 and 183.

- **consider the need for an expansion of the range**, or a change in the interval, of the independent variable

- **consider major sources of error in the investigation**. These are discussed on pp. 87–88. You should be able to judge how much these might have affected the reliability of the results.

Drawing conclusions

Again, this builds on what you did at AS, where you were also required to draw conclusions from data. Now, though, you may also have statistical analyses to help you to do this. Remember that you will be expected to use the data provided, or that you have calculated, to support your conclusion.

You may also be asked to give biological explanations of the data provided. So it is important that you go into the examination with the facts and concepts in your head that you have learned throughout your AS and A level course.

> **Expert tip**
>
> During your course:
> - Take every opportunity to practise calculating and using statistical tests.
> - Make you sure you know how to calculate the number of degrees of freedom.
> - The most important part of a statistical test is deciding what you can conclude from it. Make sure you get plenty of practice in doing this during your course.
>
> In the exam:
> - If asked, show clearly how you would use tables and graphs to display your data. Show the complete headings for rows, columns and axes, including units.
> - Show every step in each calculation that you are asked to do. Even if you get the correct answer, you will not get full marks if steps are missing.
> - Be prepared to suggest which would be the best statistical test to use in a particular situation.
> - You will not normally be asked to work through an entire statistical test, but only part of it. Read the information you are given very carefully.

A level exam-style questions and answers

This practice paper comprises structured questions similar to those you will meet in the exam.

You have 2 hours to do the paper. There are 100 marks on the paper, so you can spend just over 1 minute per mark. 85 marks are for structured questions, and there is one 15-mark question that requires more extended writing. You will get a choice of one from two questions in the actual exam paper, but in this sample paper there is only one question.

See p. 91 for advice on using this practice paper.

Exemplar paper

Question 1

The flow diagram shows part of the metabolic pathway of glycolysis.

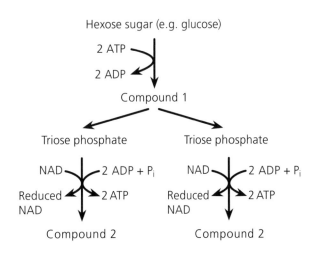

Hexose sugar (e.g. glucose)

2 ATP
2 ADP

Compound 1

Triose phosphate   Triose phosphate

NAD — 2 ADP + P$_i$ NAD — 2 ADP + P$_i$
Reduced — 2 ATP   Reduced — 2 ATP
NAD   NAD

Compound 2   Compound 2

(a) Name compound 1 and compound 2. (2 marks)
(b) State the part of the cell in which this metabolic pathway takes place. (1 mark)
(c) (i) Describe how compound 2 is converted to lactate in a human muscle cell if oxygen is not available in the cell. (2 marks)
(ii) Describe what happens to the lactate produced. (2 marks)

(Total: 7 marks)

Student A

(a) Compound 1 is fructose phosphate ✗. Compound 2 is pyruvate ✓.

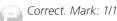 *Compound 1 is a bisphosphate, not a phosphate (it has two phosphate groups attached to it). Mark: 1/2*

Student A

(b) Cytoplasm ✓.

Correct. Mark: 1/1

Student A

(c) (i) This is anaerobic respiration. The pyruvate is changed into lactate so it does not stop glycolysis happening.

This does not answer the question. Mark: 0/2

Student A

(ii) It goes to the liver, which breaks it down ✓. This needs oxygen, which is why you breathe faster than usual when you have done a lot of exercise.

 Much more is needed for full marks. The comment about breathing rate is correct but is not relevant to this particular question. Mark: 1/2

Student B

(a) Compound 1 is hexose bisphosphate ✓. Compound 2 is pyruvate ✓.

Both are correct. Mark: 2/2

Student B

(b) Cytoplasm ✓

Correct. Mark: 1/1

Student B

(c) (i) It is combined with reduced NAD ✓, which is oxidised back to ordinary NAD. The enzyme lactate dehydrogenase ✓ makes this happen.

This is completely correct. Mark: 2/2

Student B

(ii) The lactate diffuses into the blood and is carried to the liver cells ✓. They turn it back into pyruvate again ✓, so if there is oxygen it can go into a mitochondrion and go through the Krebs cycle. Or the liver can turn it into glucose again ✓, and maybe store it as glycogen.

All correct and relevant. Mark: 2/2

Question 2

The diagram shows the structure of a chloroplast.

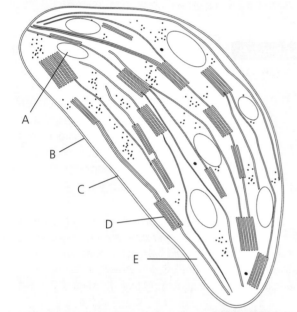

(a) Give the letter of the part of the chloroplast where each of the following takes place.
   (i) fixation of carbon dioxide   (1 mark)
   (ii) the light dependent reactions   (1 mark)
(b) A grass adapted for growing in a tropical climate was exposed to low temperatures for several days. The membranes of part D moved closer together, so that there was no longer any space between them. This prevented photophosphorylation taking place.
Explain how this would prevent the plant from synthesising carbohydrates.   (3 marks)
(c) Two groups of seedlings were grown in identical conditions for 2 weeks. One group was then grown in high-intensity light and the other group in low-intensity light, for 4 weeks. Each group of plants was then placed in containers in which carbon dioxide concentration was not a limiting factor. They were exposed to light of varying intensities and their rate of carbon dioxide uptake per unit of leaf area was measured.
The results are shown in the graph.

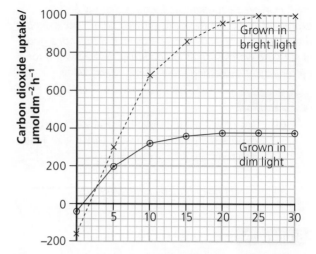

(i) Compare the effect of light intensity on the two groups of plants between 0 and $30\,mJ\,cm^{-2}\,s^{-1}$.   (4 marks)
(ii) Suggest why the plants gave out carbon dioxide at very low light intensities.   (2 marks)
(iii) Suggest two differences in the two groups of plants that could have been caused by their exposure to different intensities of light as they were growing, and that could help to explain the results shown in the graph.   (2 marks)

(Total: 13 marks)

Student A

(a) (i)   E ✓
    (ii) D ✓

Both correct. Mark: 2/2

Student A

(b) It would not be able to make any ATP ✓, which is needed for the Calvin cycle ✓. Without the Calvin cycle it would not be able to make carbohydrates.

This is correct, but lacking in detail. Mark: 2/3

Student A

(c) (i)   Neither of the groups took up any carbon dioxide when there was no light ✓ Then the quantity of carbon dioxide increased dramatically for the bright light plants, and slowly for the dim light plants. Then it levelled out, lower for the dim light plants than for the bright light ones ✓. The dim light plants levelled out at 380 and the bright light ones at $1000\,\mu mol\,dm^{-2}\,h^{-1}$ ✓.

Some good comparative points are made here. However, a fundamental error is that the answer is expressed as though the x-axis showed time — for example, the words 'slowly' and 'then' are used. It is also not a good idea to use words such as 'dramatically'. See answer B for a better way of expressing these points. Mark: 3/4

Student A

(ii) They could not photosynthesise ✓, so the carbon dioxide in their leaves just went back out into the air again.

 One correct point is made here. Mark: 1/2

Student A

(iii) The ones that had grown in the bright light could have bigger leaves and more chlorophyll. ✓

 The suggestion about bigger leaves is not correct. Even if the plants did have bigger leaves, this would not affect the results, because the carbon dioxide uptake is measured per unit area (look at the units on the y-axis of the graph). The second point is a good suggestion. Mark: 1/2

Student B

(a) (i) E ✓
(ii) D ✓

 Both are correct. Mark: 2/2

Student B

(b) No ATP would be made ✓ in the light dependent reaction, so there would not be any available for the light independent reactions, where carbohydrates (triose phosphate) are made in the Calvin cycle ✓. ATP is needed to convert GP to triose phosphate ✓ (along with reduced NADP) and also to help regenerate RuBP ✓ from the triose phosphate so the cycle can continue.

 All correct and with good detail. Mark: 3/3

Student B

(c) (i) Below about 0.5 light intensity, both groups gave out carbon dioxide ✓. As light intensity increased, the amount of carbon dioxide taken up by the plants grown in bright light increased more steeply ✓ than for the ones grown in dim light. In the group grown in dim light, the maximum rate of carbon dioxide uptake was $380\,\mu mol\,dm^{-2}\,h^{-1}$, whereas for the ones in bright light it was much higher ✓, at $1000\,\mu mol\,dm^{-2}\,h^{-1}$ ✓. For the bright light plants, the maximum rate of photosynthesis was not reached until the light intensity was $25\,mJ\,cm^{-2}\,s^{-1}$, but for the ones grown in dim light the maximum rate was reached at a lower ✓ light intensity of $20\,mJ\,cm^{-2}\,s^{-1}$.

 This is good answer, with some comparative figures quoted (with their units). Five credit-worthy points have been made, although there are only 4 marks available for this question. Note the avoidance of any vocabulary that could be associated with time. Mark: 4/4

Student B

(ii) When the light intensity was very low, the plants would not be able to photosynthesise so they would not take up any carbon dioxide ✓. However, they would still be respiring (they respire all the time) so their leaf cells would be producing carbon dioxide ✓, which would diffuse out into the air. Normally, the cells would take up this carbon dioxide for photosynthesis.

 This is a good answer. Mark: 2/2

Student B

(iii) The plants grown in the light would probably be a darker green because they would have more chlorophyll ✓ in their chloroplasts, so they would be able to absorb more light and photosynthesise faster. They might also have more chloroplasts in each palisade cell. ✓ And leaves sometimes produce an extra layer of palisade cells if they are in bright light. ✓

 This answer actually contains three correct points, and two of them have been explained, which was not required. The student could have got 2 marks with a much shorter answer. Nevertheless, this answer shows good understanding of the underlying biology. Mark: 2/2

Question 3

A group of islands contains three species of mice, each species being found on only one island. A fourth species is found on the mainland. A region of the DNA of each species was sequenced, and the percentage differences between the samples were calculated. The results are shown in the table.

	Mainland	Island A	Island B
Island A	6.1		
Island B	4.8	9.7	
Island C	5.2	10.3	7.5

(a) Discuss how these results suggest that the species of mouse on each island has evolved from the species on the mainland, and not from one of the other island species. **(5 marks)**

(b) Each of these four species of mice is unable to breed with any of the other species, even if they are placed together.

Suggest how reproductive isolation between the mice could have arisen, and explain its role in speciation. (5 marks)

(Total: 10 marks)

Student A

(a) The three island mice each have DNA more similar to the mainland mouse than to each other ✓✓. So they have probably all evolved from the mainland mouse. If one of the island mice had evolved from another island mouse, their DNA would be more similar. ✓

ⓔ *This answer shows that the student has managed to work out what the table shows, but it does not provide enough detail to get all of the marks available. Mark: 3/5*

Student A

(b) If they are on different islands, they will have different selection pressures ✓ so the mice might end up different. They might have different courtship behaviour ✓, so they will not be able to mate with each other ✓. You have to get reproductive isolation to produce a new species.

ⓔ *This is a reasonable description. The last sentence is moving towards another mark but it really only repeats what is already in the question. Mark: 3/5*

Student B

(a) From the table, we can see that each island mouse's DNA is more similar to the DNA of the mainland mouse than to any of the other island mice. ✓ For example, the island C mice have DNA that is 10.3% different from the island A mice and 7.5% different from the island B mice, but only 5.2% different from the mainland mice. ✓ The longer ago two species split away from each other, the more different we would expect their DNA to be. This is because the longer the time, the more mutations ✓ are likely to have occurred, so there will be different base sequences ✓ in the DNA.

ⓔ *This is a good answer to a difficult question. Good use is made of the data in the table, and some of the figures are used to support the answer. The last part of the answer explains why differences in DNA base sequence indicate degree of relationship. Mark: 5/5*

Student B

(b) A species is defined as a group of organisms that can interbreed with each other to produce fertile offspring. ✓ So to get new species you have to have something that stops them reproducing together so genes cannot flow ✓ from one species to the other. This might happen if different selection pressures ✓ acted on two populations of a species, so that different alleles were selected for ✓ and over many generations their genomes became more different ✓. So they might be the wrong size and shape ✓ to be able to breed with each other.

ⓔ *The answer begins with a clearly explained link between speciation and reproductive isolation, and then goes on to describe how two populations could become reproductively isolated. Mark: 5/5*

Question 4

Gibberellin is a plant hormone (plant growth regulator) that is involved in stem elongation and seed germination.
(a) Explain why plants with the genotype lele **are dwarf, whereas plant with genotype** LeLe **or** Lele **can grow tall.** (4 marks)
(b) During the germination of a wheat seed, gibberellin activates the production of the enzyme amylase.
   (i) Explain how it does this. (4 marks)
   (ii) Outline the role of amylase in seed germination. (2 marks)

(Total: 10 marks)

Student A

(a) **Le** is dominant to **le**, so the heterozygote **Lele** can make gibberellin, but **lele** cannot ✓.

ⓔ *This is correct, but there is no explanation of how allele **Le** or **le** are involved in the production of gibberellin. The answer does not make any link between the production of gibberellin and the ability to grow tall. Mark: 1/4*

Student A

(b) (i)   Gibberellin (GA) is made by the seed when it is warm and it takes up water. The GA breaks the dormancy of the seed. The GA makes the cells in the seed secrete amylase.

ⓔ *Once again, the student has written nothing wrong, but there is no description of how gibberellin brings about amylase synthesis. Mark: 0/4*

Student A

(ii)   Amylase breaks down starch to maltose. ✓

ⓔ *This is correct but it is not enough for an answer to an A level question. Mark: 1/2*

Student B

(a) The gene **Le/le** codes for the synthesis of an enzyme that is needed to make gibberellin. ✓ **Le** codes for a functioning enzyme, but **le** does not. ✓ **Le** is dominant and **le** is recessive, so the plant only needs one **Le** allele to be able to make the enzyme. ✓ Gibberellin is needed to make plants grow tall because it simulates stem elongation ✓, so plants with the genotype **lele** do not make any gibberellin ✓ and do not grow tall.

ⓔ *This is a good answer. It explains each step of the process by which the allele **Le** allows plants to grow tall. Mark: 4/4*

Student B

(b) (i) Before the seed starts to germinate, proteins called DELLA proteins are bound to transcription factors inside the cells in the seed. ✓ When a seed takes up water, this stimulates the synthesis of GA by the tissues of the embryo plant. GA combines with a receptor ✓ inside the cells in the seed, and also with an enzyme. This activates the enzyme ✓ so that it can break down the DELLA protein ✓. The transcription factor is now free to bind with DNA ✓ in the cell, which initiates the transcription of genes for amylase ✓.

ⓔ *Here is another excellent answer. The series of events is described well and good use is made of technical language, such as 'synthesis' and 'transcription factor'. Mark: 4/4*

Student B

(ii) The amylase hydrolyses starch stored in the endosperm, producing maltose ✓, which is soluble and is transported to the embryo for use as an energy resource ✓ and also as raw material from which cellulose can be synthesised to make new cell walls ✓.

ⓔ *This answer has plenty of relevant detail, and scores maximum marks. The student tells us what amylase does (again making good use of technical language) and how this is important for growth of the embryo. Mark: 2/2*

Question 5

The Irish Threatened Plant Genebank was set up in 1994 with the aim of collecting and storing seeds from Ireland's rare and endangered plant species. The natural habitat of many of these species is under threat. For each species represented in the bank, seeds are separated into active and base collections. The active collection contains seeds that are available for immediate use, which could be for reintroduction into the wild, or for germination to produce new plants and therefore new seeds. The base collection is left untouched. Some of the base collection is kept in Ireland, and some is kept at seed banks in other parts of the world.

(a) Suggest why seed banks separate stored seeds into active and base collections. (2 marks)

In 2001, an investigation was carried out into the effect of long-term storage on the ability of the seeds to germinate. Fifteen species were tested. In each case, 100 seeds were tested. It was not

possible to use more because in many cases this was the largest number that could be spared from the seed bank. In most cases, the germination rate of the seeds had already been tested when they were first collected in 1994, so a comparison was possible with the germination rates in 2001 after 7 years of storage.

The graphs show the results for two species, *Asparagus officinalis* and *Sanguisorba officinalis*.

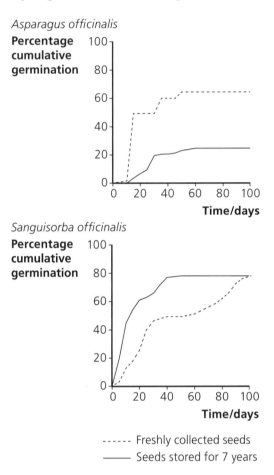

Asparagus officinalis

Sanguisorba officinalis

------ Freshly collected seeds
——— Seeds stored for 7 years

(b) (i) Compare the germination rates of stored and fresh seeds of *Sanguisorba officinalis*. (3 marks)

(ii) Compare the effect of storage on the germination rates of *Sanguisorba officinalis* and *Asparagus officinalis*. (3 marks)

(c) It has been suggested that species stored as seeds in seed banks have different selection pressures acting on them compared with the same species living in the wild.

(i) Explain why the selection pressures in a seed bank and in the wild are likely to be different. (2 marks)

(ii) Suggest how the possible harmful effects of these differences could be minimised. (5 marks)

(Total: 15 marks)

Student A

(a) So they always have some spare.

This is not enough for a mark. Mark: 0/2

Student A

(b) (i) The stored ones germinated better than the fresh ones. The stored ones go up more quickly than the fresh ones. But they all end up at the same place, about 80%. ✓

This answer loses out by poor wording, and not being clear enough about exactly what is being described. The word 'better' in the first sentence could mean that the seeds germinated more quickly, or that more of them germinated, so this needs to be clarified. 'Go up more quickly' is also not clearly related to germination. The last sentence is rather generously awarded a mark for the idea that eventually about 80% of the seeds in each batch germinated. Mark: 1/3

Student A

(ii) The *Sanguisorba* seeds germinated better when they had been stored, but the *Asparagus* seeds germinated better when they were fresh. Storing the *Sanguisorba* seeds made them germinate faster, but the *Asparagus* seeds germinated slower. ✓

Again, poor wording means that this answer only gets 1 mark. The word 'better' is used in the first sentence and, as explained above, this is not a good word to use in this context. One mark is given for the idea that storage caused slower germination in Asparagus *but faster germination in* Sanguisorba. *Mark: 1/3*

Student A

(c) (i) The conditions in which the seeds grow might be different in the seed bank from in the wild.

This is not quite correct because seeds do not grow. Growth only happens after germination, so it is seedlings and plants that grow. The student needs to think more carefully about what happens to seeds in a seed bank. Mark: 0/2

Student A

(ii) The seeds could be grown in conditions like those in the wild.

Again, this has not been clearly thought out. Mark: 0/5

Student B

(a) Having active collections is good because it means there are seeds available that can be used for something. But you must always keep some seeds in storage, because the whole point of a seed bank is that it stores seeds and these need to be kept safe so they do not get destroyed. ✓ If all of them got used for growing plants, then perhaps the plants would die ✓ and you would not have any seeds left.

This is not very well expressed, but the right ideas are there. Mark: 2/2

Student B

(b) (i) The stored seeds germinated much faster ✓ than the fresh ones. By 10 days, about 60% of the stored seeds had germinated, but only about 8% ✓ of the fresh ones. By 50 days, all of the stored seeds that were going to germinate (about 80%) had germinated. It took 100 days for 80% of the fresh seeds to germinate. ✓ We cannot tell if any more would have germinated after that because the line is still going up when the graph stops. ✓

Clear comparative points have been made about the speed at which germination happened, and also about the maximum percentage of seeds that germinated. Mark: 3/3

Student B

(ii) Storage seemed to help the germination for *Sanguisorba*, but it made it worse for *Asparagus*. For *Asparagus*, storage made it germinate slower, but it germinated faster for *Sanguisorba*. ✓ And for *Asparagus* only about 25% of the stored seeds germinated compared with 60% of the fresh seeds ✓, but with *Sanguisorba* about the same percentage of seeds germinated for both fresh and stored ✓.

This is a good answer, which again makes clear comparisons and discusses both the speed of germination and the percentage of seeds that eventually germinated. Mark: 3/3

Student B

(c) (i) In the seed bank, the seeds are just stored. So the ones that survive are the ones that are best at surviving in those conditions as dormant seeds. ✓ In the wild, the plants have to be adapted to grow in their habitat ✓, so maybe they have to have long roots or big leaves or whatever. In the seed bank, that does not matter.

This does answer the question, but it could perhaps have been written a little bit more carefully and kept shorter. Mark: 2/2

Student B

(ii) You might get seeds that are really good at surviving in a seed bank but when they germinate they produce plants that are not very good at surviving in the wild. To avoid this, you could keep on collecting fresh seeds from plants in the wild ✓, and only storing them for a little while before replacing them with new ones ✓. If you had to store seeds for a long time, you could keep germinating some of them ✓ and growing them in conditions like in the wild ✓ and then collect fresh seeds from the ones that grew best ✓.

This is a good answer to a tricky question. The student has made several sensible suggestions, including storing the seeds for a shorter time and periodically exposing the plants to natural conditions where the 'normal' selection pressures will operate. Mark: 5/5

Question 6

Two parents with normal skin and hair colouring had six children, of whom three were albino. Albino people have no colouring in their skin or hair, due to having an inactive form of the enzyme tyrosinase. Tyrosinase is essential for the formation of the brown pigment melanin.

(a) The normal allele of the tyrosinase gene is A, and the allele that produces faulty tyrosinase is a.
State the genotypes of the parents and their albino children. (2 marks)

(b) Albinism is a relatively frequent condition in humans, but one of these albino children had a very unusual phenotype. While most of her hair was white, the hair of her eyebrows developed some brown colouring, as did the hair on her hands and lower legs. Genetic analysis suggested that a mutation had occurred in the faulty tyrosinase allele.
Suggest why it is likely that this mutation occurred in the ovaries or testes of the girl's parents, rather than in her own body.
(2 marks)

(c) The graph opposite shows how the activity of normal tyrosinase and tyrosinase taken from the albino girl were affected by temperature.
(i) Compare the effects of temperature on normal tyrosinase and the albino girl's tyrosinase. (3 marks)

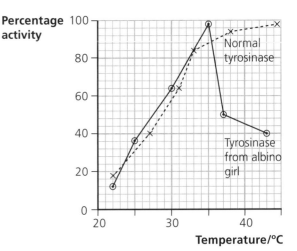

(ii) Studies on the production and activity of tyrosinase in living cells found that normal tyrosinase leaves the endoplasmic reticulum shortly after it has been made, and accumulates in vesicles in the cytoplasm where it becomes active. However, the tyrosinase in the albino girl's cells only did this at temperatures of 31 °C or below. At 37 °C, her tyrosinase accumulated in the endoplasmic reticulum. Use your answer to (i) and the information above to discuss explanations for the distribution of colour in the hair on different parts of the albino girl's body.
(4 marks)
(Total: 11 marks)

Student A

(a) **Aa** and **aa** ✓✓

It is not clear which genotype refers to the parents and which to the children, but the answer has generously been given the benefit of the doubt. Mark: 2/2

Student A

(b) If it had been in one of her own cells, then only that cell would be affected and not the whole body. ✓

This is correct, as far as it goes. Mark: 1/2

Student A

(c) (i) Both enzymes get more active as temperature increases. ✓ The normal enzyme is still getting more active even at 45 °C. ✓ The girl's enzyme has an optimum activity at 35 °C ✓ and above that it quickly ✗ gets a lot less active.

The answer begins well with a sentence about both enzymes. Two clear distinguishing points are then made. However, the use of the term 'quickly' is inappropriate as there is nothing about time on the graph. 'Steeply' would have been better. Mark: 3/3

Student A

(ii) Maybe her legs and eyebrows and hands were colder ✓, like in a Himalayan rabbit. So the enzyme would not work well where it was hotter ✓ and would not make melanin.

This is a difficult question, and the student has done well to think of what he/she had learned about the interaction between genes and environment in animal hair colour and recognise that it could relate to this situation. However, the answer does not refer clearly to the information provided, and this needs to be done in order to achieve more marks. Mark: 2/4

Student B

(a) Parents **Aa** and **Aa**. ✓ Albino children **aa**. ✓

This is correct and clear. Mark: 2/2

Student B

(b) If it had occurred in her own body, then it would probably be in only one cell ✓ so you would not see the effects in cells in different parts of the body. If it had been in one of the parents, then the mutated gene would have been in the zygote ✓, so when the zygote divided by mitosis the gene would get copied ✓ into every cell in her body.

This is absolutely correct. Mark: 2/2

Student B

(c) (i) The activity of both enzymes increases as temperature increases from 22 °C to 35 °C. ✓ However, above that the activity of the girl's enzyme drops really steeply ✓ so it has only 50% activity at 37 °C. But the normal enzyme keeps on increasing its activity up to where the graph ends at 45 °C. ✓ At 38 °C it has 95% activity, almost twice as much as for the girl's enzyme. ✓

This is a good answer, which mentions particular points on the curves and provides a quantitative comparison between the figures for the two enzymes at a particular temperature (38 °C). Mark: 3/3

Student B

(ii) The extremities of the girl's body will be cooler than other parts ✓, so the tyrosinase will be able to work here because according to the graph it can work well up to 35 °C ✓. It will also be able get out of the endoplasmic reticulum ✓ in her cells and into the

Student B continued

vesicles where it becomes active. This is why the hairs on her hands and legs are darker because the enzyme was able to make melanin there. ✓ But in warmer places the enzyme stops working and cannot get out of the RER, so no melanin is made. ✓

This is an excellent answer. Both sets of information (from the graph and the description about the endoplasmic reticulum) are used to provide an explanation linked to what the student had already learned about the control of hair colour in other situations. Mark: 4/4

Question 7

(a) Describe how gel electrophoresis can be used to separate DNA fragments of different lengths.

(6 marks)

(b) The diagram shows the results of electrophoresis on DNA samples taken from a mother, her child and its alleged father.

Explain what can be concluded from these results. **(3 marks)**

(Total: 9 marks)

Student A

(a) First you cut the DNA up into pieces using restriction enzymes. ✓ Then you put the DNA onto some agarose gel ✓ in a tank and switch on the power supply so the DNA gets pulled along the gel ✓. The bigger pieces move more slowly, so they end up not so far along the gel as the smaller pieces. ✓ You cannot see the DNA so you need to stain it with something so it shows up.

There are no errors in this answer, but a little more detail is needed. Mark: 4/6

Student A

(b) The child has one band that is in its mother's DNA and another that is in the alleged father's DNA. ✓ So he could be the father. ✓

Two correct points are made. Mark: 2/3

Student B

(a) The DNA is cut into fragments using restriction enzymes ✓, which cut it at particular base sequences. Then you place samples of the DNA into little wells in agarose gel ✓ in an electrophoresis tank. A voltage is then applied ✓ across the gel. The DNA pieces have a small negative charge so they steadily move towards the positive ✓ terminal. The larger they are, the more slowly they move ✓ so the smaller ones travel further ✓ than the big ones. After a time, the power is switched off so the DNA stops moving. You can tell where it is by using radioactivity ✓, so the DNA shows up as bands on a photographic film.

All the points made are correct and there is enough detail for full marks. Mark: 6/6

Student B

(b) The top band for the child matches the top band for the mother ✓, and the bottom band for the child matches the bottom band for the father ✓. So the alleged father could be the child's father ✓, though we cannot be 100% certain of that because there could be another man who has this band as well ✓.

This is a clear and thorough conclusion. Mark: 3/3

Question 8

Adrenaline is a hormone secreted by the adrenal glands. One of its effects is to cause the hydrolysis of glycogen to glucose in liver cells.

(a) Explain what is meant by *hydrolysis*. (2 marks)

(b) Complete the passage, which describes how adrenaline brings about the hydrolysis of glycogen to glucose in a liver cell.
Adrenaline binds to a receptor molecule in the cell .... This activates a .... messenger called cyclic .... , which binds to .... proteins in the cytoplasm. This causes phosphate groups to be added to enzymes, which activates them. The activation of these enzymes results in the activation of many molecules of glycogen .... , which hydrolyse glycogen to glucose. (5 marks)

(c) Explain how a single molecule of adrenaline can cause the production of a very large number of glucose molecules. (3 marks)

(Total: 10 marks)

Student A

(a) The breakdown of a large molecule to a small one. ✓

This is correct, but only one point has been made. Mark: 1/2

Student A

(b) Adrenaline binds to a receptor molecule in the cell **membrane**. ✓ This activates a **hormone** ✗ messenger called cyclic **ADP** ✗, which binds to **kinase** ✓ proteins in the cytoplasm. This causes phosphate groups to be added to enzymes, which activates them. The activation of these enzymes results in the activation of many molecules of glycogen **hydrolase** ✗, which hydrolyse glycogen to glucose.

Two words are correct. This kind of question is often thought of as being easy, but that is far from the case — you need to choose the words carefully. Mark: 2/5

Student A

(c) One enzyme can break down a lot of molecules of glycogen, because enzymes are not used up. ✓

One correct point is made. Mark: 1/3

Student B

(a) The breakdown of a large molecule by the addition of water. ✓ In glycogen, hydrolysis breaks the glycosidic bonds ✓ between glucose molecules.

This is correct. Mark: 2/2

Student B

(b) Adrenaline binds to a receptor molecule in the cell **membrane** ✓. This activates a **second** ✓ messenger called cyclic **AMP** ✓, which binds to **kinase** ✓ proteins in the cytoplasm. This causes phosphate groups to be added to enzymes, which activates them. The activation of these enzymes results in the activation of many molecules of glycogen **phosphorylase** ✓, which hydrolyse glycogen to glucose.

All the words are correct. Mark: 5/5

Student B

(c) When one adrenaline molecule binds to the cell surface membrane, it sets off an enzyme cascade. ✓ At each step, one enzyme causes the activation of many more enzymes, so by the time you get to glycogen phosphorylase there are many of them activated. ✓ Moreover, one glycogen phosphorylase molecule can cause the breakdown of many glycogen molecules ✓, because enzymes are not destroyed in the reaction. Each glycogen molecule is a polymer of glucose, so breaking down just one glycogen molecule produces a very large number of glucose molecules. ✓

This is a good answer. The student has thought of several different reasons (enzyme cascade, the ability of one enzyme molecule to catalyse many reactions, and the fact that one glycogen molecule contains many glucose molecules). Many students would think of just one reason and stop there, so this student has done very well. Mark: 3/3

Question 9

(a) Describe how a nerve impulse crosses a
    cholinergic synapse. (9 marks)
(b) Outline the functions of a sensory neurone and
    a motor neurone in a reflex arc. (6 marks)

(Total: 15 marks)

This is the extended writing question.

Student A

(a) In order for the action potential to cross the synaptic cleft,
it has to go through a process.

A nerve impulse arrives at a synapse in the form of an action
potential. It causes sodium and calcium gates to open so
these ions ✓ go into the presynaptic membrane ✗. The
calcium causes the vesicles in the presynaptic neurone to
fuse with the membrane ✓ and release acetycholine into
the synapse. This binds to the receptors on the postsynaptic
membrane ✓ and this makes them pump ✗ sodium ions into
the neurone which starts up another action potential in the
second neurone, because it depolarises it ✓.

Then the enzyme acetylcholinesterase causes the
acetylcholine to go back into the presynaptic neurone ✗, so it
returns to the resting potential again.

*This answer is not clear, and the examiner cannot always
be certain what the student means. For example, the
calcium ions go through the presynaptic membrane and into
the neurone, not 'into' the membrane. Notice that no mark is
given for the mention of calcium until the term 'ions' is used
later on. There is an error in the next to last sentence; the
sodium ions are not pumped into the postsynaptic neurone,
but pass through the opened sodium ion channels by
diffusion, down their concentration gradient. The last sentence
also contains an error, as acetycholinesterase breaks down
acetycholine. In any case, this process is not strictly relevant to
a description of how the impulse crosses the synapse, because
it takes place after that event has finished. Overall, this is not a
strong answer, with lack of detail and several errors. Mark: 4/9*

Student A

(b) The sensory neurone transmits an action potential from
a receptor into the spinal cord. ✓ The action potential
crosses a synapse and then passes along a motor neurone
to an effector, such as muscle. ✓ This makes the muscle
contract. ✓ This is a reflex action. The muscle automatically
contracts without you having to think about it.

*There are three correct statements here, but a good
IGCSE student could have given this answer and there is
not sufficient detail for a high mark at A level. Mark: 3/6*

Student B

(a) When the action potential arrives at the presynaptic
knob ✓, it causes calcium ion channels to open ✓. Ca^{2+}
flood into the neurone ✓, down their concentration
gradient ✓. The knob contains many tiny vesicles full of the
neurotransmitter ✓ acetylcholine ✓. The calcium ions make
these vesicles move to the presynaptic membrane and
fuse with it ✓, releasing the acetylcholine into the synaptic
cleft ✓.

This cleft is very small, so it takes only a millisecond or two
for the acetylcholine to diffuse ✓ across it. On the other side
of the cleft, there are receptor molecules in the postsynaptic
membrane, and the acetylcholine molecules fit perfectly into
these. ✓ This makes sodium ion channels in the postsynaptic
membrane open ✓, so sodium ions flood in down their
concentration gradient ✓. This depolarises the membrane ✓
(gives it a positive charge inside), which sets up an action
potential in the postsynaptic neurone.

*This is a good answer. Although it is short, it is packed
with correct and relevant detail. There are no mistakes,
and no important steps have been omitted. The extended
answer questions usually have many more marking points
than the total number of marks available, so even though this
answer has 13 ticks it can still only get the maximum mark.
Mark: 9/9*

Student B

(b) A sensory neurone has its cell body in the ganglion in the
dorsal root of a spinal nerve. It has a very long dendron
that carries action potentials from a receptor towards its
cell body ✓, and a shorter axon that carries the action
potentials into the spinal cord (or brain) ✓. The ending of
the dendron may be within a specialised receptor such
as a Pacinian corpuscle in the skin. ✓ Pressure acting on
the Pacinian corpuscle depolarises the membrane of the
dendron and generates an action potential. ✓ (In general,
receptors transfer energy from a stimulus into energy in an
action potential.)

The motor neurone has its cell body within the central
nervous system (in the brain or the spinal cord). It has many
short dendrites and a long axon. It will have many synapses,
including several with sensory neurones. ✓ Thus the action
potential from a sensory neurone can cross the synapse and
set up an action potential in the motor neurone ✓, which will
then transmit it to an effector such as a muscle or gland ✓.
The action potential then causes the effector to respond, for
example by contracting (if it is a muscle). ✓

In a reflex arc, the impulses travel directly from the sensory
to the motor neurone (or sometimes via an intermediate
neurone between them) without having to be processed
in the brain. ✓ This means the pathway from receptor to
effector is as short as possible, so the response can happen
very quickly.

*This is a good answer, taking each neurone in turn and
concentrating on functions, with only brief references to
structure where these are directly relevant to function. Mark:
6/6*

Now test yourself answers

Topic 1

1 79 mm = 79 000 µm

actual size of object = $\dfrac{\text{size of image}}{\text{magnification}}$

$= \dfrac{79\,000}{16\,000}$

$= 4.9\,\mu m$

2 100 graticule units = 8 × 0.01 mm

$= 0.08\,mm$

So 1 graticule unit $= \dfrac{0.08}{100}$

$= 0.0008\,mm$

$= 0.0008 \times 1000\,\mu m$

$= 0.8\,\mu m$

Cell measures 84 graticule units

$= \dfrac{84}{0.8} = 105\,\mu m$

3 From scale bar with the virus diagram, 5 mm represents 1 nm

Diameter of virus in diagram = 40 mm

So actual diameter of virus $= \dfrac{40}{5} = 8\,nm$

From scale bar with the bacterium diagram, 12 mm = 0.1 µm = 1 × 10⁵ nm

Length of bacterium in diagram = 54 mm

So actual length $= \dfrac{54}{12} \times 10^5 = 4.5 \times 10^5\,nm$

Therefore $\dfrac{4.5 \times 10^5}{8} = 5.6 \times 10^4$ viruses could line up inside it.

Topic 2

1 Fructose: monosaccharide; gives positive result with Benedict's test

Glucose: monosaccharide; gives positive result with Benedict's test

Lactose and maltose: disaccharides; lactose is formed from glucose and galactose, maltose is formed from glucose and glucose; give positive result with Benedict's test

Sucrose: disaccharide; formed from glucose and fructose; does not give positive result with Benedict's test

2 Cellulose is made up long chains of glucose linked by β 1–4 glycosidic bonds. This results in the glucose molecules being alternately one way up, then the other. This means that chains lying side by side can form hydrogen bonds with each other, making microfibrils. Starch, on the other hand, is made up of long chains of glucose linked by α 1–4 glycosidic bonds. It twists into a spiral, and one starch molecule does not form bonds with others, so no fibrils are formed. Moreover, it is difficult to break down β 1–4 glycosidic bonds, so not many organisms can digest cell walls.

3 First do the reducing sugar test. Use excess Benedict's solution so that all the reducing sugar is used up in the test. Filter to remove the brick red precipitate. Collect the filtrate. It would be a good idea to do the reducing sugar test again, just to check that you really have got rid of all the reducing sugar. If you have, then do the non-reducing sugar test.

4 a Secondary structure is the first-level coiling or folding of the polypeptide chain. This is often an alpha helix or a beta pleated strand. Tertiary structure is the three-dimensional folding of the polypeptide chain, for example to form a globular shape.

b A collagen molecule is a chain of amino acids linked by peptide bonds. A collagen fibre is made up of many collagen molecules linked together by covalent bonds between lysine molecules.

5 a A hydrogen bond is a weak attraction between a small negative charge (e.g. on the oxygen of an −OH group) and a small positive charge (e.g. on the hydrogen of an −OH group).

b Hydrogen bonds cause forces of attraction between water molecules. This means that extra energy has to be applied to separate the molecules from one another. This means that the melting point and boiling point are higher than for similar compounds that do not have hydrogen bonding between their molecules.

c Carbohydrates, proteins

Topic 3

1 Variable to change (independent variable): catalase from different fruits.

Variable to measure (dependent variable): initial rate of reaction (e.g. volume of oxygen given off per second)

Variables to control: volume and concentration of catalase extract; volume and concentration of substrate (hydrogen peroxide solution); temperature; pH

- Use a liquidiser to mash up the same mass of two fruits. Run the liquidiser at the same speed for the same period of time.
- Filter the extracts and collect equal volumes of each.
- Measure equal volumes of the extract into several test tubes.
- Use serial dilution to make up a range of concentrations of hydrogen peroxide (see p. 78 for how to do this).
- Add a small amount of pH 7 buffer to each tube.
- Place all tubes (fruit extract and substrate) in a water bath at 37 °C and leave for 20 minutes to come to temperature.
- Add fruit extract to one of the hydrogen peroxide tubes, and measure the rate of oxygen production (e.g. by collecting over water, or by placing on a balance and measuring loss of mass).
- Repeat for each hydrogen peroxide concentration.
- Plot graphs of oxygen production against substrate concentration for each tube.
- Use your graphs to calculate the initial rate of reaction for each tube (see p. 28 for how to do this).
- Plot graphs of initial rate of reaction against substrate concentration for each of the two fruit juice extracts.
- Use your graphs to find V_{max} for each fruit juice extract.

- Read off the substrate concentration that corresponds to $\frac{1}{2}V_{max}$. This gives you the value of K_m.
- The fruit juice extract with the larger value of K_m is the one with greater affinity for its substrate.

2 a Temperature, enzyme concentration, substrate concentration, competitive inhibitors

   b Temperature, pH and non-competitive inhibitors

3 Enzymes do not contaminate the final product; enzymes are more stable so can work at a wider range of temperatures and pH; enzymes are not lost, so can be reused many times.

Topic 4

1 Diffusion takes place when molecules or ions can move freely through the membrane, through the phospholipid bilayer. Facilitated diffusion takes place when molecules or ions can only move through protein channels in the membrane, not through the phospholipid bilayer. Note that both processes are passive, and movement is down a concentration gradient.

2 Facilitated diffusion takes place down a concentration gradient, and is passive. Active transport takes place up a concentration gradient and requires energy input from the cell. Note that in both of these processes, molecules or ions move through proteins in the membrane.

Topic 6

1 There is only room in the DNA double helix for one nucleotide with one ring, and one with two rings, to link together. Moreover, C and G join with three hydrogen bonds, while A and T join with two.

2 Three

3 GUA, CCU, GAC

Topic 7

1 Your plan diagram should:
- be large — preferably larger than the diagram in the book
- be drawn with clear, clean lines
- not show any individual cells
- be made up of four concentric lines — a pair of lines quite close together on the outside to represent the epidermis and root hairs; a line much closer to the centre of your drawing representing the boundary between the cortex and the endodermis; another line close to the last one representing the inner edge of the endodermis; and then a cross-shaped structure representing the xylem and phloem.

2 a Plant has to provide energy.

   b Plant does not have to provide energy.

3 Movement of water up the xylem is down a water potential gradient, from roots to leaves. This gradient is maintained by transpiration in the leaves, which lowers the water potential there. The gradient is always in the same direction.

Movement of sap in phloem is down a pressure gradient, from source to sink. High pressure is produced by active loading of sucrose into the phloem at a source, which causes water to follow by osmosis. Different parts of the plant can act as sources at different times — for example leaves when they photosynthesise or roots when the starch stored in them is broken down. Different parts can also act as sinks at different times — for example roots in autumn when they are building up starch stores or flowers when they are using sugars to produce nectar (so the flow can be upwards or downwards).

4 a Leaves are sources; flowers and roots are sinks.

   b Leaves are sources; fruits and roots are sinks.

   c Roots are sources; leaves are sinks.

   d Roots are sources; leaves and flowers are sinks.

Topic 8

1 No. The pulmonary artery carries deoxygenated blood.

2 Time taken for one complete heart beat is about 0.75 seconds. So in one minute there will be 60/0.75 = 80 beats.

Topic 10

1 The cause of an infectious disease is the pathogen that enters the body and makes you ill. A vector for a disease is an organism that transmits the pathogen into the body.

2 HIV-positive means that the person has been infected with the virus. AIDS means that the virus is causing symptoms of disease.

Topic 11

1 The mitochondria provide ATP for protein synthesis and secretion. The rough endoplasmic reticulum provides sites for the synthesis of antibodies (which are proteins). The Golgi bodies prepare the antibodies for secretion.

Topic 12

1 Phosphorylation raises the energy level of the molecule, making it able to take part in the reaction.

2 One proton and one electron

3 If there is no soda lime then no carbon dioxide will be absorbed. If the organisms are respiring aerobically using glucose as a substrate, then the volume of oxygen taken in will equal the volume of carbon dioxide given out, and there will be no change in the level of fluid in the manometers.

Topic 13

1 The absorption spectrum shows the wavelengths of light that can be absorbed. The action spectrum shows which wavelengths of light can be used. We would therefore expect the two to be very similar to one another.

2 The energy ends up in ATP molecules.

3 Substrate: carbon dioxide

Product: intermediate compound *or* GP

4 From the light dependent reaction.

Topic 14

1 No matter what the environmental temperature, enzymes will always be able to work at close to their optimum temperature. This means that metabolism can continue at a fairly steady rate during day and night, and in different seasons, and in different climatic regions of the world. This means that animals can be active at all of these times and in different places.

2 One signalling molecule (e.g. a hormone molecule) activates large numbers of glycogen phosphorylase enzymes.

3 The biosensor gives a digital readout, which is easy to interpret and has a continuous range. Dip sticks give a colour that you have to match against a colour scale, and have a discontinuous range.

4 The loss of water from the filtrate increases the concentration of all the solutes in it.

5 There would be a water potential gradient from the cytoplasm of the red blood cell into the plasma. Water would move out of the cell by osmosis, through its partially permeable cell surface membrane. This would decrease the volume of the cell, so it would shrink.

Topic 15

1 The opening and closing of the potassium ion channels is caused by changes in the potential difference across the membrane. They open when this is positive inside, and close when it is negative inside. They are said to be voltage-gated channels.

2 The vesicles containing transmitter substance are present only in the presynaptic neurone, so the impulses can only be transmitted from this one, not from the postsynaptic neurone. Also, there are only receptors for the transmitter substance on the postsynaptic neurone, not on the presynaptic neurone.

3 A muscle fibre is made up of many myofibrils arranged side by side. It is a specialised cell. It is surrounded by a cell surface membrane (sarcolemma) and contains several nuclei and other organelles, especially mitochondria and endoplasmic reticulum. A myofibril is one of several cylindrical structures within a muscle fibre, and contains filaments. A filament can be one of two types — either actin or myosin — and is essentially a protein molecule.

4 The mitochondria are the sites in which the Krebs cycle and the electron transport chain generate ATP. ATP is required for the detachment of the myosin heads from the actin filaments, allowing the filaments to slide past one another and cause the muscle to contract.

5 Your table could include the following:
- Both are small molecules that act as cell signalling first messengers.
- Animal hormones are carried in the blood, but plant hormones travel in phloem or by diffusion.
- Animal hormones are made in endocrine glands, but plant hormones are made in tissues that are not organised into glands.

6 A plant with genotype **LeLe** already makes gibberellin, so can already grow tall. Adding more gibberellin is unlikely to have any effect.

Topic 16

1 The gametes in the self-pollinated plant will both be made from mother cells of the same genotype. These gametes will vary slightly in their genotypes because of independent assortment and crossing over, but this can only arise from reshuffling of the same set of alleles in each. The gametes in the cross-pollinated plant will have come from mother cells of different genotypes, in different plants. They will therefore have a different range of alleles. This will result in a wider range of possible allele combinations in the offspring from the cross-pollinated plant.

2 Parents' phenotypes Blood group AB × Blood group AB

Parents' genotypes I^AI^B I^AI^B

Gametes' genotypes I^A I^B I^A I^B

Offspring genotypes and phenotypes

	I^A	I^B
I^A	I^AI^A Blood group A	I^AI^B Blood group AB
I^B	I^AI^B Blood group AB	I^BI^B Blood group B

We would therefore expect to get blood group A, B and AB in the ratio 1:1:2.

3 The allele for colour blindness is carried on the X chromosome. A man passes only his Y chromosome to his son.

4 Parents' phenotypes

brown hair, long legs × black hair, short legs

Parents' genotypes **AaLl** **aall**

Gametes' genotypes AL Al aL al al

Offspring genotypes and phenotypes

	AL	Al	aL	al
al	AaLl brown, long	Aall brown, short	aaLl black, long	aall black, short

5 Parents' phenotypes

brown eyes, dark fur × brown eyes, dark fur

Parents' genotypes

EeFf   **EeFf**

Gametes' genotypes

(EF) (Ef) (eF) (ef)   (EF) (Ef) (eF) (ef)

Offspring genotypes and phenotypes

	(EF)	(Ef)	(eF)	(ef)
(EF)	EEFF brown, dark	EEFf brown, dark	EeFF brown, dark	EeFf brown, dark
(Ef)	EEFf brown, dark	EEff brown, pale	EeFf brown, dark	Eeff brown, pale
(eF)	EeFF brown, dark	EeFf brown, dark	eeFF blue, dark	eeFf blue, dark
(ef)	EeFf brown, dark	Eeff brown, pale	eeFf blue, dark	eeff blue, pale

The shaded rows and columns show the uncommon gametes and offspring phenotypes. We would therefore expect most offspring to be in the ratio 3 brown, dark:1 blue, pale. There will be a small number of other combinations.

6 a If the genes are linked:

Parents' phenotypes   yellow, large × yellow, large

Parents' genotypes   **YyPp**   **YyPp**

Gametes' genotypes   (YP) (yp)   (YP) (yp)

Offspring genotypes and phenotypes

	(YP)	(yp)
(YP)	**YYPP** yellow, large	**YyPp** yellow, large
(yp)	**YyPp** yellow, large	**yypp** white, small

This would give a ratio of 3 yellow, large:1 white, small if the genes are totally linked. If there is some crossing over, we would expect to get small numbers of other phenotypes.

b If the genes are not linked:

Parents' phenotypes

yellow, large   ×   yellow, large

Parents' genotypes

YyPp   **YyPp**

Gametes' genotypes

(YP) (Yp) (yP) (yp)   (YP) (Yp) (yP) (yp)

Offspring genotypes and phenotypes

	(YP)	(Yp)	(yP)	(yp)
(YP)	YYPP yellow, large	YYPp yellow, large	YyPP yellow, large	YyPp yellow, large
(Yp)	YYPp yellow, large	YYpp yellow, small	YyPp yellow, large	Yypp yellow, small
(yP)	YyPP yellow, large	YyPp yellow, large	yyPP white, large	yyPp white, large
(yp)	YyPp yellow, large	Yypp yellow, small	yyPp white, large	yypp white, small

This would give a ratio of 9 yellow, large:3 yellow, small: 3 white, large:1 white, small

7 First, work out the expected numbers if there is no linkage. You should find that you would expect a 1:1:1:1 ratio of phenotypes. (See answer to Question 4 above.)

	White, long	White, short	Grey, long	Grey, short
Observed numbers, O	12	2	2	12
Expected numbers, E	7	7	7	7
$O - E$	5	−5	−5	5
$(O - E)^2$	25	25	25	25
$\dfrac{(O - E)^2}{E}$	3.6	3.6	3.6	3.6
$\Sigma \dfrac{(O - E)^2}{E} = 14.4$				

We have four different categories, so the number of degrees of freedom is 3. Looking up the value of chi-squared for 3 degrees of freedom, we find that our number lies between a probability of 0.01 and 0.001 that the null hypothesis (i.e. there is no linkage) is correct. This is much less than the critical value of 0.05, so we can say that the null hypothesis is not correct, and that our results show that the two genes are linked.

8 Parents' phenotypes   agouti   ×   agouti

Parents' genotypes   **AaBB**   **AaBb**

Gametes' genotypes   (AB) (aB)   (AB) (Ab) (aB) (ab)

Offspring genotypes and phenotypes

	(AB)	(aB)
(AB)	AABB agouti	AaBB agouti
(Ab)	AABb agouti	AaBb agouti
(aB)	AaBB agouti	aaBB black
(ab)	AaBb agouti	aaBb black

We would expect offspring in the ratio 3 agouti : 1 black.

9 A white mouse could have the genotype **AAbb**, **Aabb** or **aabb**. We could cross this mouse with a pure-breeding black mouse, with the genotype **aaBB**.

If the unknown mouse has the genotype **AAbb**, then all the offspring will be agouti (**AaBb**).

If the unknown mouse has the genotype **Aabb**, then half the offspring will be agouti (**AaBb**) and half will be black (**aaBb**).

If the unknown mouse has the genotype **aabb**, then all the offspring will be black (**aaBb**).

10 It is wasteful to produce the enzyme when there is no substrate for it. The bacterium saves energy and materials by not making the enzyme in these circumstances.

Topic 17

1 The graph should have two peaks, like a two-humped camel.

2 Frequency of the homozygous recessive animals (q^2) is 100 in 5000, or 0.02.

  a   So q is the square root of 0.02, which is 0.14.

  b p is 1 − 0.14, which is 0.86.

      So the frequency of heterozygotes is $2pq$, which is 2 × 0.14 × 0.86 = 0.24.

      We would therefore expect about 24% of the animals to be heterozygotes.

3 Courtship behaviour and sperm survival are pre-zygotic. Chromosome number and inability to form gametes are post-zygotic.

Topic 18

1 Estimated total population size $= \dfrac{38 \times 40}{6}$

                        $= 253$

2 Calculate n/N for each species. Then square each value. Add them up and subtract from 1.

Species	n	n/N	$(n/N)^2$
A	2	0.010	0.000
B	35	0.174	0.030
C	1	0.005	0.000
D	81	0.403	0.162
E	63	0.313	0.098
F	2	0.010	0.000
G	5	0.025	0.001
H	11	0.055	0.003
I	1	0.005	0.000

$\Sigma\left(\dfrac{n}{N}\right)^2 = 0.294$

So $D = 1 - 0.294 = 0.706$

Topic 19

1 Selective breeding can only reshuffle alleles that are already present in a species. Gene technology introduces completely new alleles from a different species.

2 Gene therapy can only add genes, not remove them. If a condition is caused by a dominant allele, this will still be in the cell, so even if a 'correct' recessive allele is introduced, it will not have any effect.

A level experimental skills and investigations

1

x	$(x - \bar{X})$	$(x - \bar{X})^2$
23.0	−17.5	306.25
46.5	6.0	36.00
52.0	11.5	132.25
32.5	−8.0	64.00
43.0	2.5	6.25
50.5	10.0	100.00
26.5	−14.0	196.00
31.5	−9.0	81.00
54.0	13.5	182.25
28.0	−12.5	156.25
47.5	7.0	49.00
39.0	−1.5	2.25
58.5	18.0	324.00
47.5	7.0	49.00
33.5	−7.0	49.00
42.5	2.0	4.00
26.5	−14.0	196.00
51.0	10.5	110.25
38.0	−2.5	6.25
38.5	−2.0	4.00
mean $\bar{X}$ 40.5		$\Sigma(x - \bar{X})^2 = 2051.75$

So $s = \sqrt{\dfrac{2051.75}{19}} = 10.4$

2 $S_M = \dfrac{s}{\sqrt{n}}$

$= 10.4/4.47$

$= 2.3\,\text{mm}$

3

Quadrat	Number of species P	Rank for species P	Number of species Q	Rank for species Q
1	6	6	5	7
2	12	2=	22	2
3	1	9	3	9=
4	0	10	4	8
5	16	1	35	1
6	8	5	10	5
7	11	4	11	4
8	4	8	9	6
9	5	7	3	9=
10	12	2=	21	3

Quadrat	Rank for species P	Rank for species Q	Difference in rank, D	D^2
1	6	7	−1	1
2	2=	2	0	0
3	9	9=	0	0
4	10	8	2	4
5	1	1	0	0
6	5	5	0	0
7	4	4	0	0
8	8	6	2	4
9	7	9=	−2	4
10	2=	3	−1	1
				$\Sigma D^2 = 14$

Substitute into the equation:

$r_s = 1 - \dfrac{6 \times 14}{10^3 - 10}$

$\quad = 1 - \dfrac{84}{990}$

$\quad = 1 - 0.08$

$\quad = 0.92$

This is well above the critical value for r_s when $n = 10$, so there is a significant correlation between the distributions of P and Q.

4

Mollusc	Shell length/ mm (x)	Body mass/ g (y)	xy
1	38	6.1	231.8
2	18	3.6	64.8
3	20	3.2	64.0
4	31	5.7	176.7
5	12	2.6	31.2
6	12	3.2	38.4
7	25	4.7	117.5
8	20	3.6	72.0
9	29	4.7	136.3
10	19	3.3	62.7

mean

$\bar{x} = 22.4$

$\bar{y} = 4.07$

$\Sigma xy = 995.40$

$n\bar{x}\bar{y} = 10 \times 22.4 \times 4.07 = 911.68$

standard deviation for $x = 8.34$

standard deviation for $y = 1.17$

$r = \dfrac{\Sigma xy - n\bar{x}\bar{y}}{ns_x s_y}$

So:

$r = \dfrac{995.40 - 911.68}{97.578}$

$= 0.86$

This number is close to 1, suggesting that there is a linear correlation between these two sets of data.